THIS BOOK BELONGS TO

_____

_____

_____

To join our mailing list and see other titles available

Website: www.captaintimpublishing.com

Email: info@captaintimpublishing.com

Missing numbers

Fill in the missing numbers below

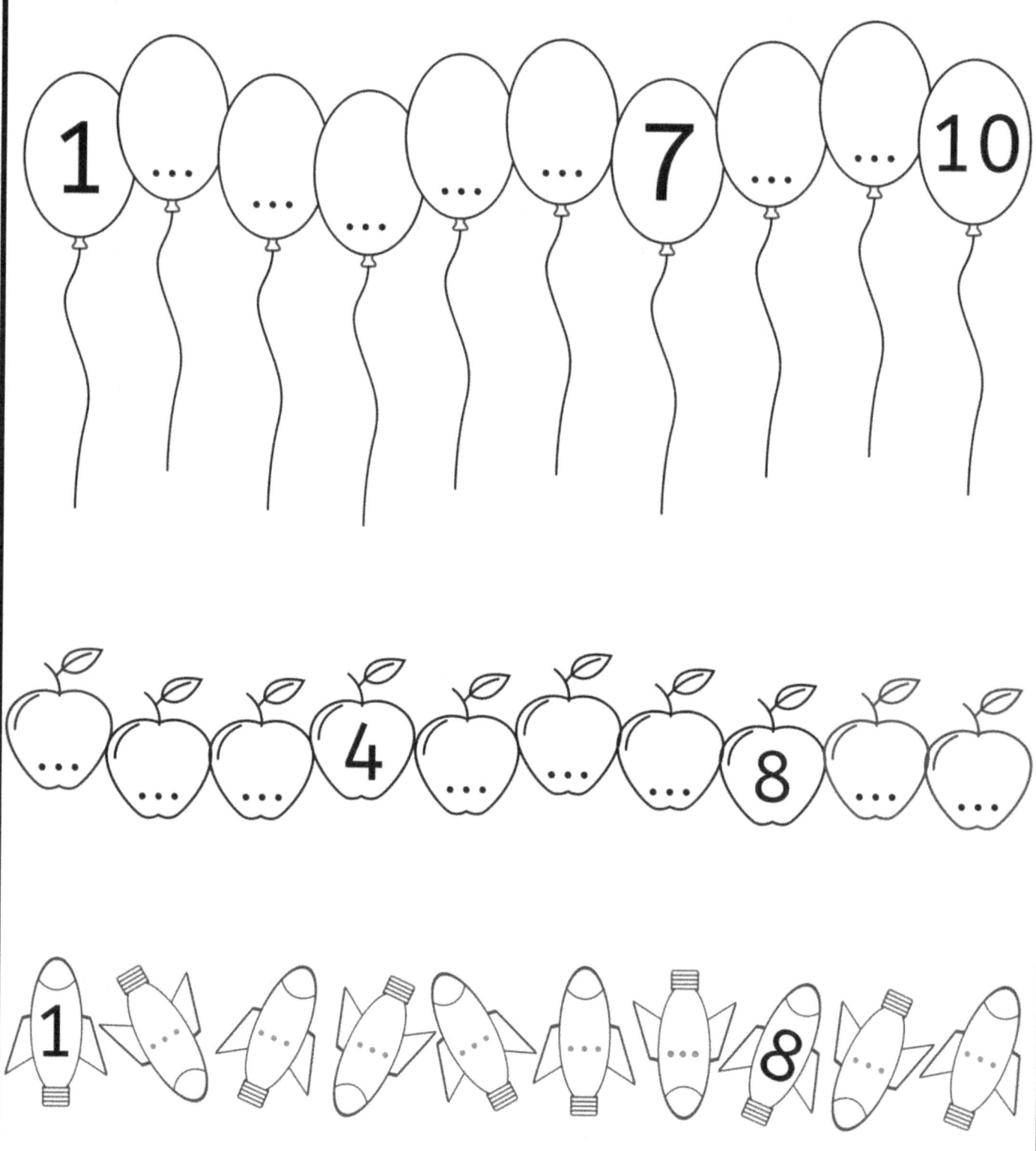

Counting

Count and circle the correct answer

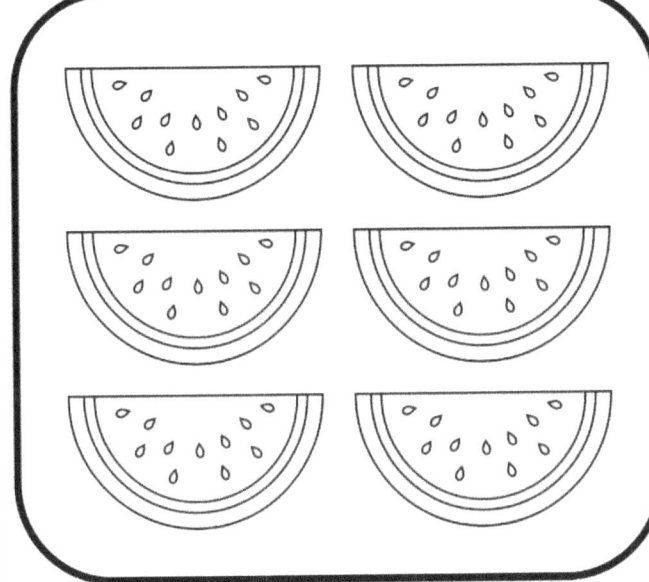

| 2 | 6 | 8 |

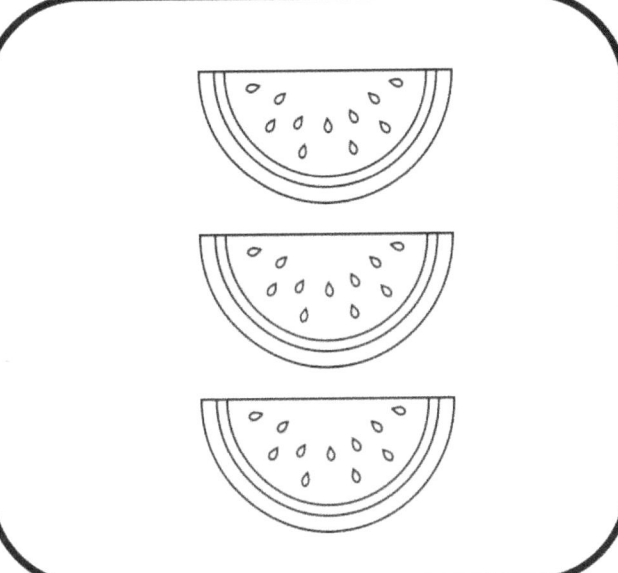

| 3 | 1 | 2 |

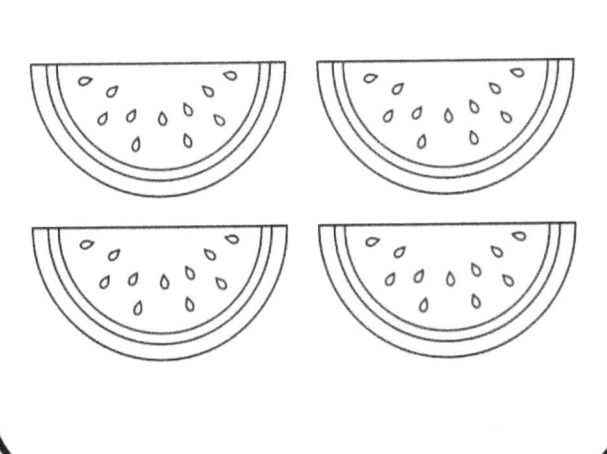

| 8 | 7 | 4 |

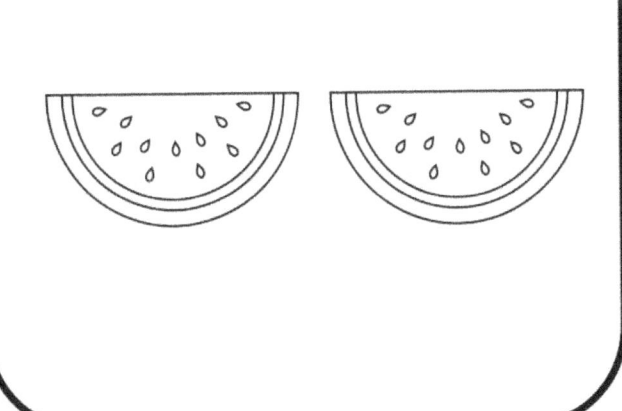

| 2 | 3 | 5 |

Addition

Add and write the answers in the
boxes then color the penguins that equal 5

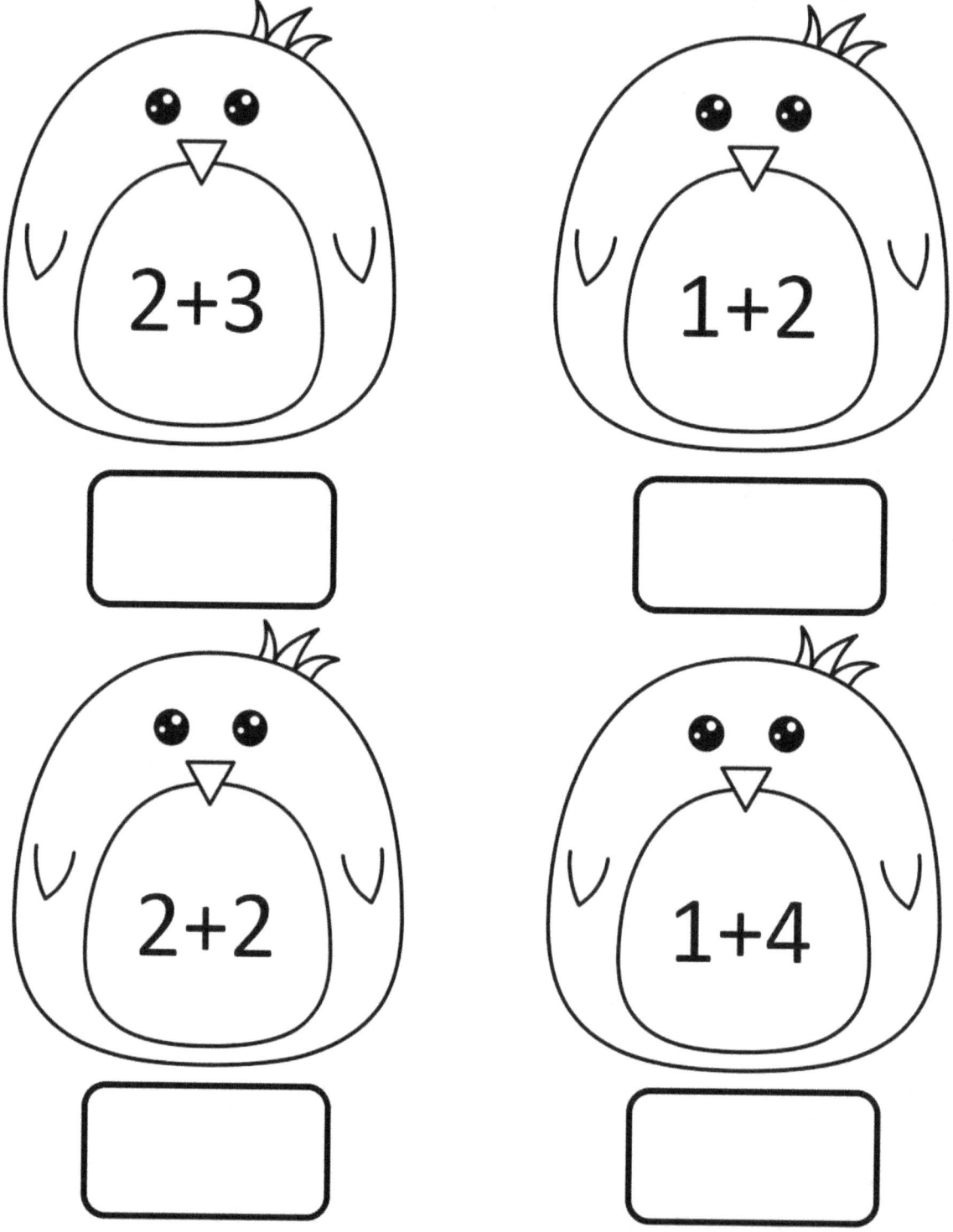

Tracing numbers

Trace the number and count items in each group.
Color the pictures that have 1 item in the set

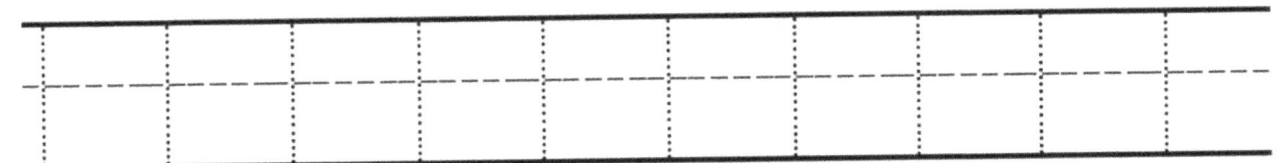

More, less or equal

Choose the correct answer

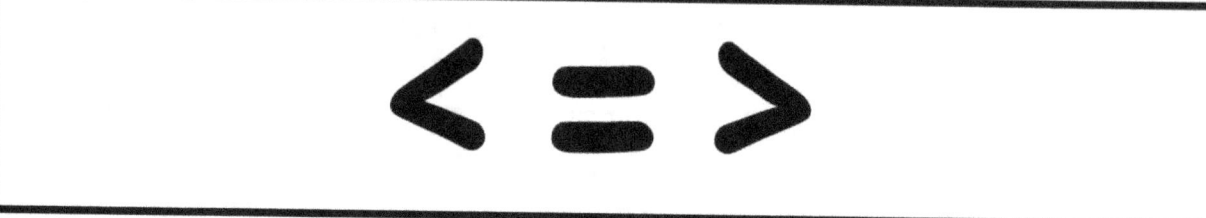

12		12
10		3
3		2

Addition

Count the rainbows and fill in the boxes

$\boxed{1}$ + 1 = $\boxed{2}$

$\boxed{}$ + 1 = $\boxed{}$

$\boxed{}$ + 1 = $\boxed{}$

$\boxed{}$ + 1 = $\boxed{}$

Subtraction

Subtract and write your answer in the box

3 - 1 = ☐

4 - 2 = ☐

2 - 2 = ☐

1 - 0 = ☐

2 - 1 = ☐

Subtraction

Subtract and write your answers in the box then color any watermelons that equal 3

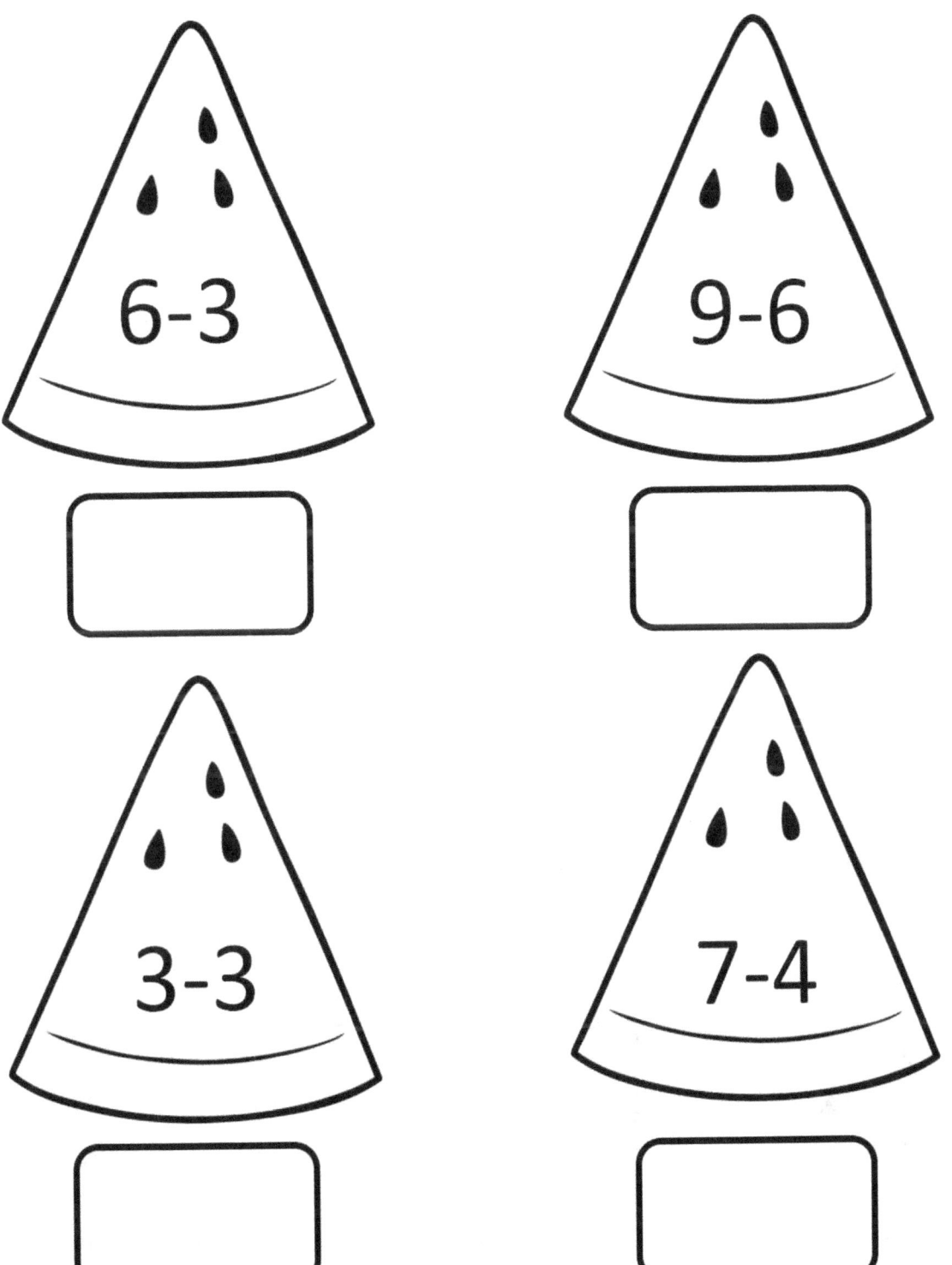

More, less or equal

Choose the correct answer

Number maze

Color the number 1 through the maze to the stop sign

Matching

Add and connect dots with the correct answer

5 + 3 ●	● 6
4 + 2 ●	● 4
6 + 4 ●	● 9
2 + 7 ●	● 5
5 + 2 ●	● 8
2 + 3 ●	● 7
1 + 2 ●	● 10
4 + 0 ●	● 3

Subtraction

Subtract the numbers from the middle number

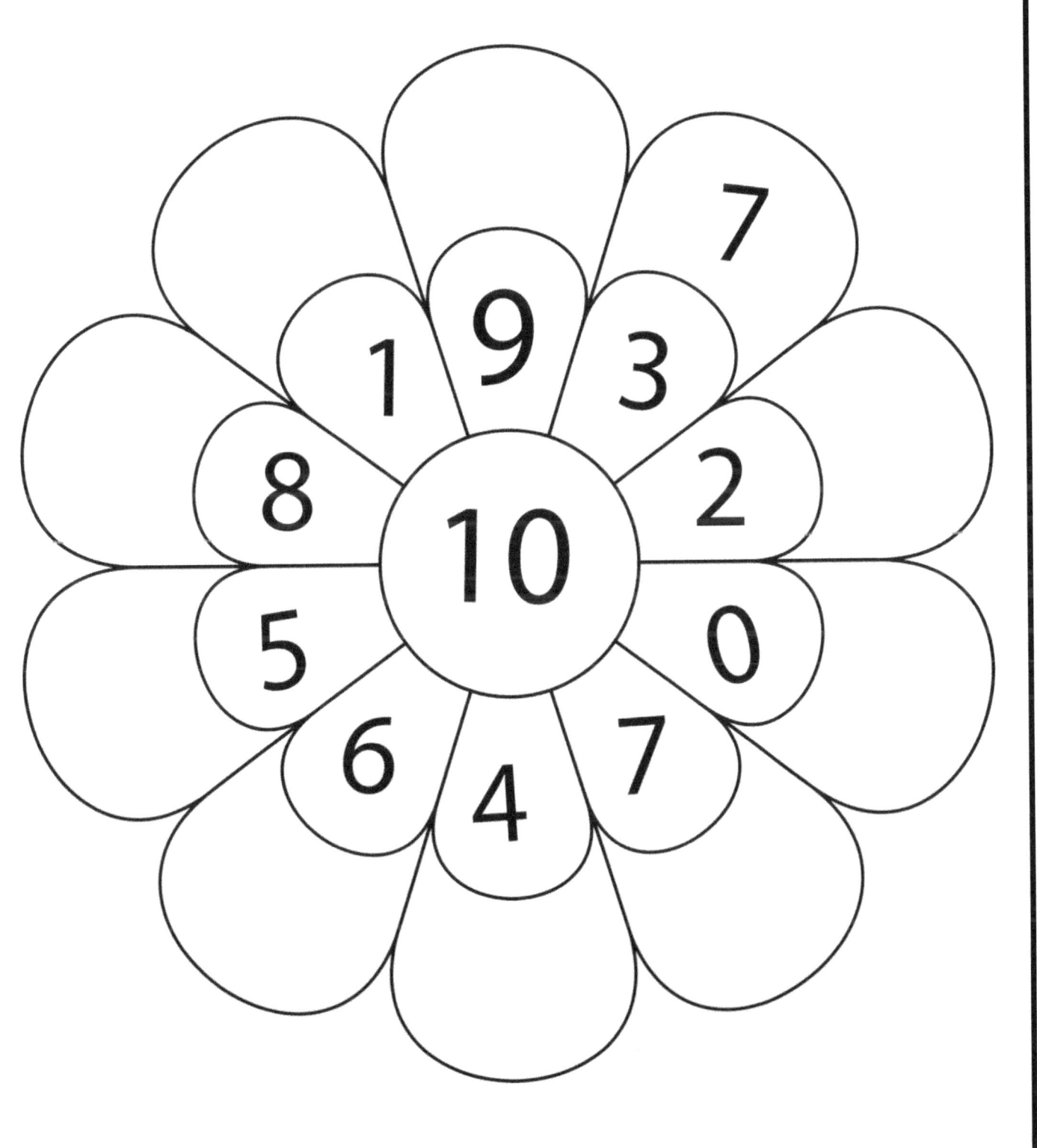

Missing numbers

Fill in the missing numbers below

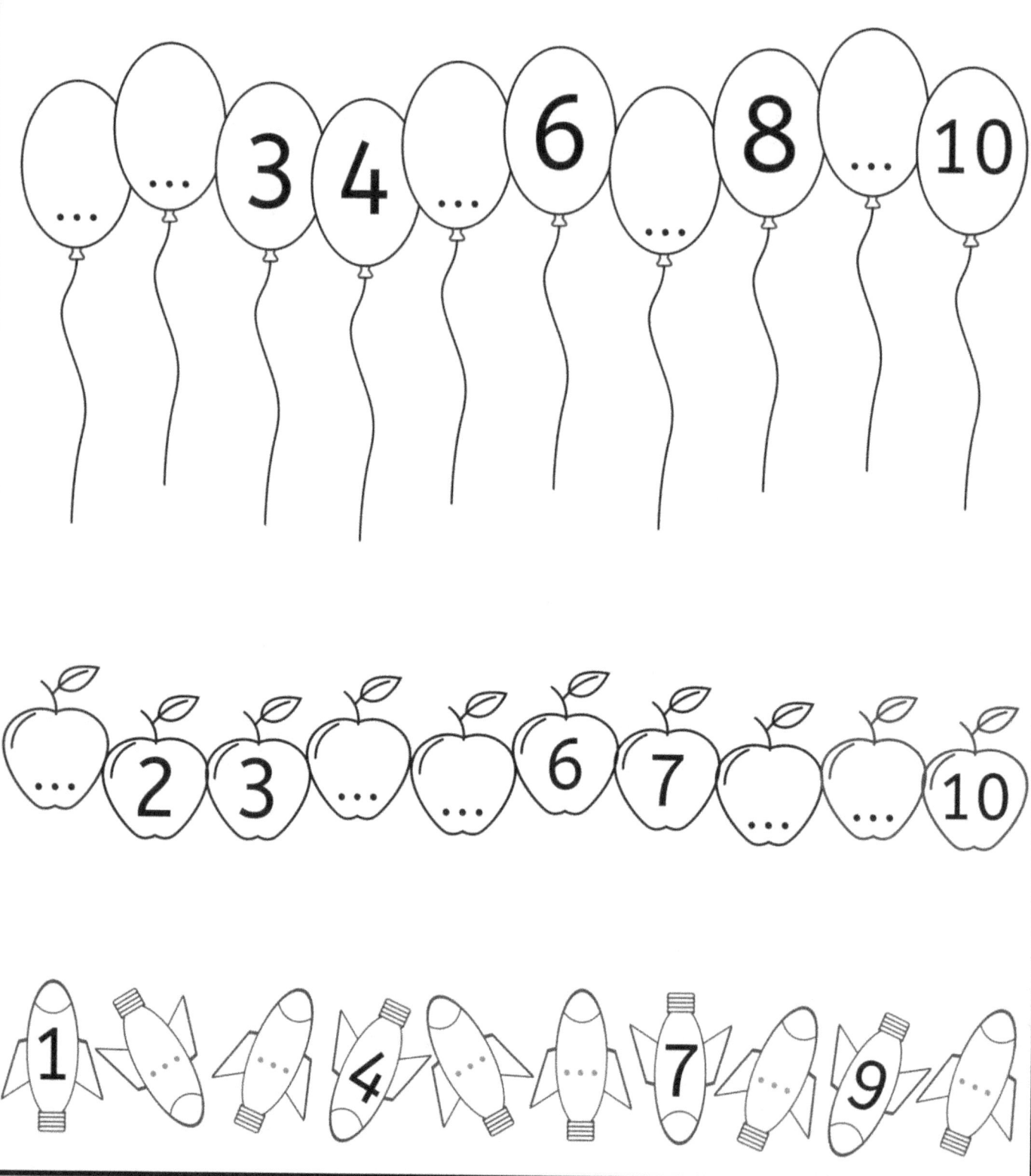

Counting

Count and circle the correct answer

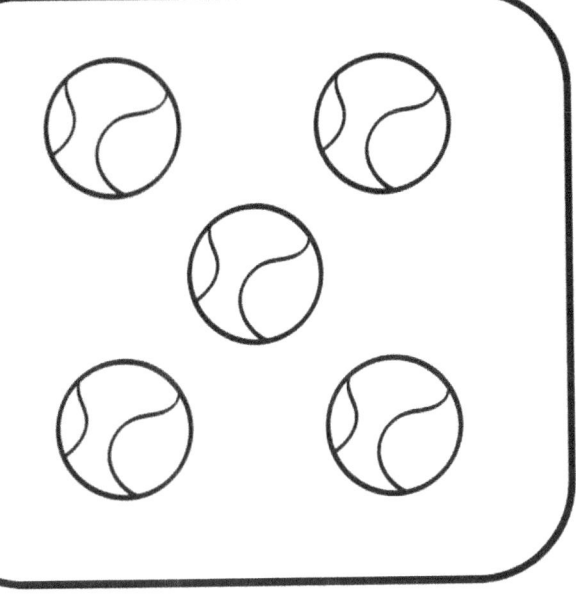

| 3 | 5 | 1 |

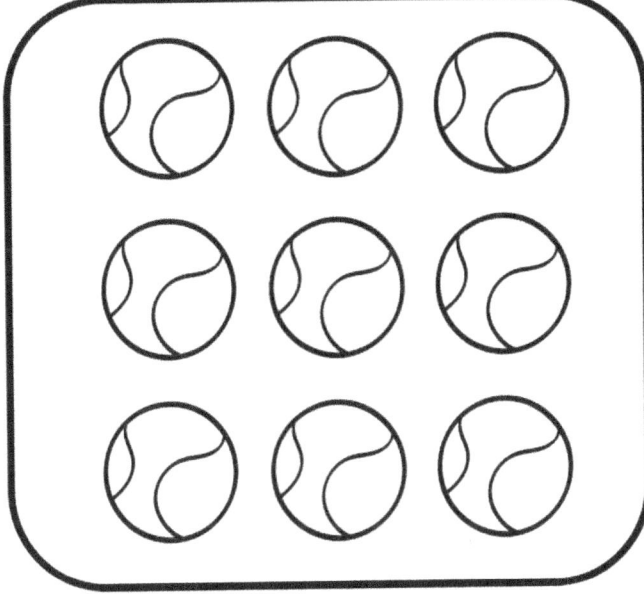

| 6 | 7 | 9 |

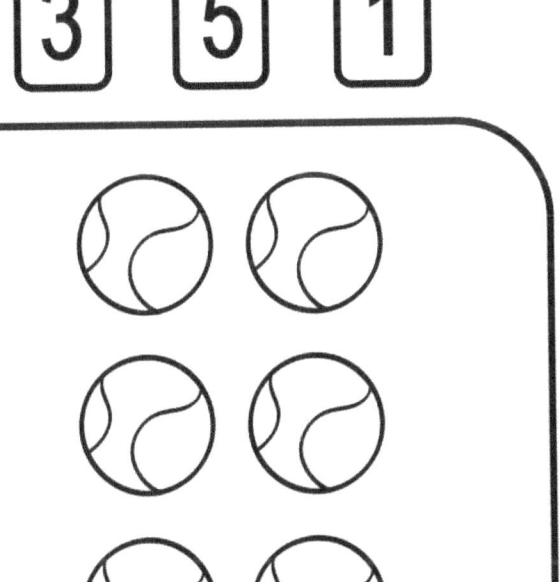

| 2 | 6 | 5 |

| 7 | 2 | 6 |

Addition

Add and write your answers in the
box then color any clouds that equal 3

Tracing numbers

Trace the number and count items in each group.
Color the pictures that have 2 items in the set

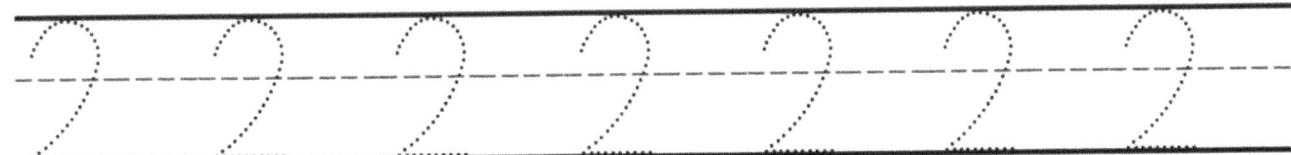

Addition

Count the butterflies and fill in the boxes

5 + 2 = 7   ☐ + 2 = ☐

☐ + 2 = ☐   ☐ + 2 = ☐

Number maze

Color the number 2 through the maze to the stop sign

Tell the time

Draw two hands on the
clock face to show the time

 11:15

 5 :45

 6 :20

 2 :50

 9 :10

 7 :40

12:55

3 :30

 8 :05

 4 :35

 1 :25

 10:00

Subtraction

Subtract and write your
answer in the box

10 - 5 = ☐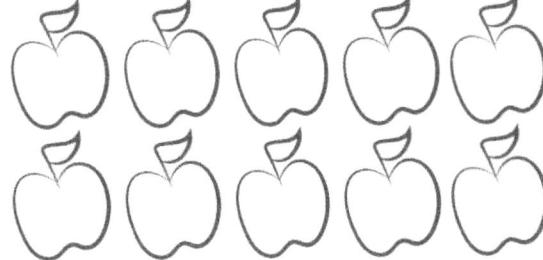

8 - 7 = ☐

9 - 5 = ☐

7 - 2 = ☐

4 - 1 = ☐

Subtraction

Subtract and write your answers in the box then color any bears that equal 3

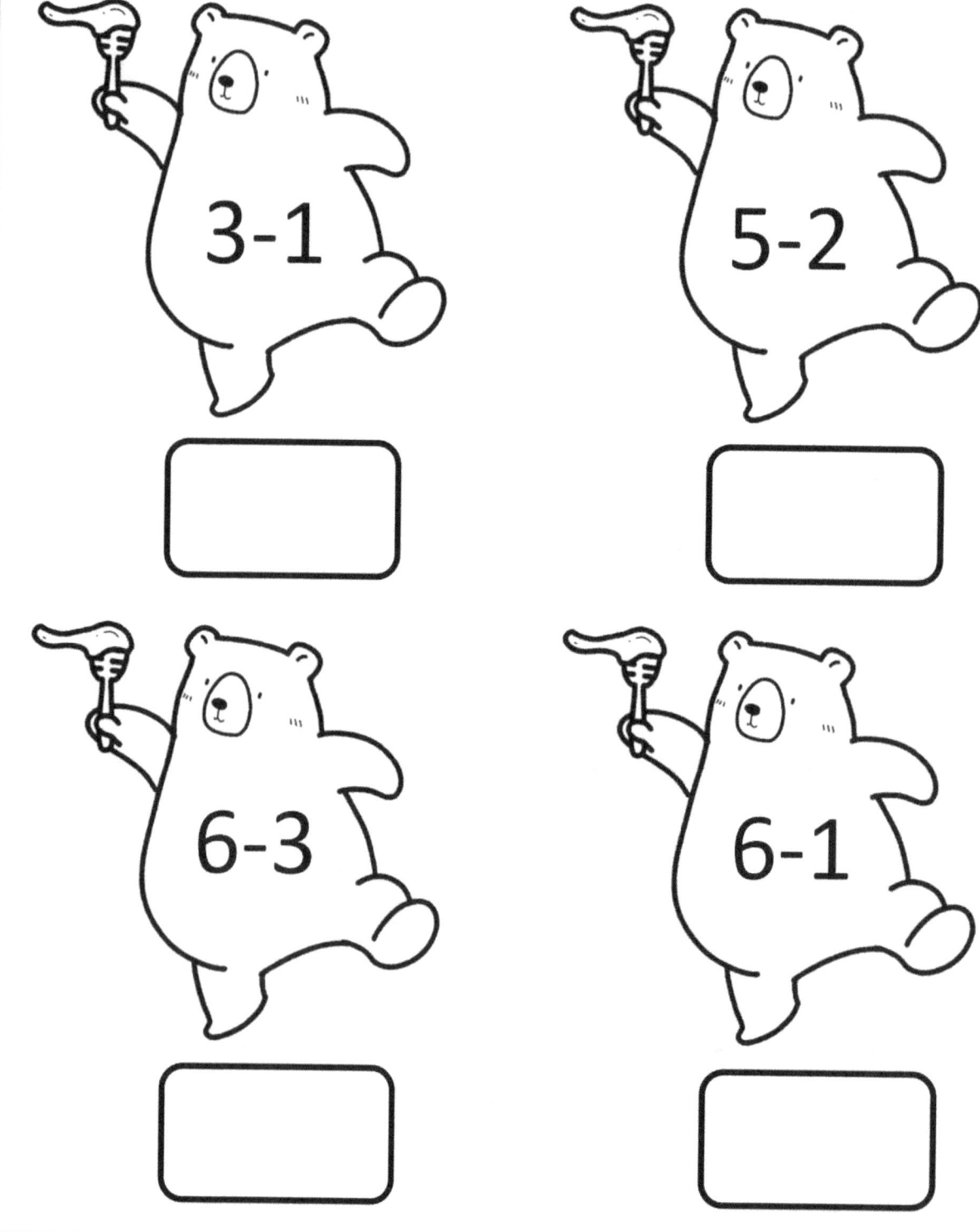

Number maze

Color the number 3 through the maze to the stop sign

Counting

Count the objects and write the answers below

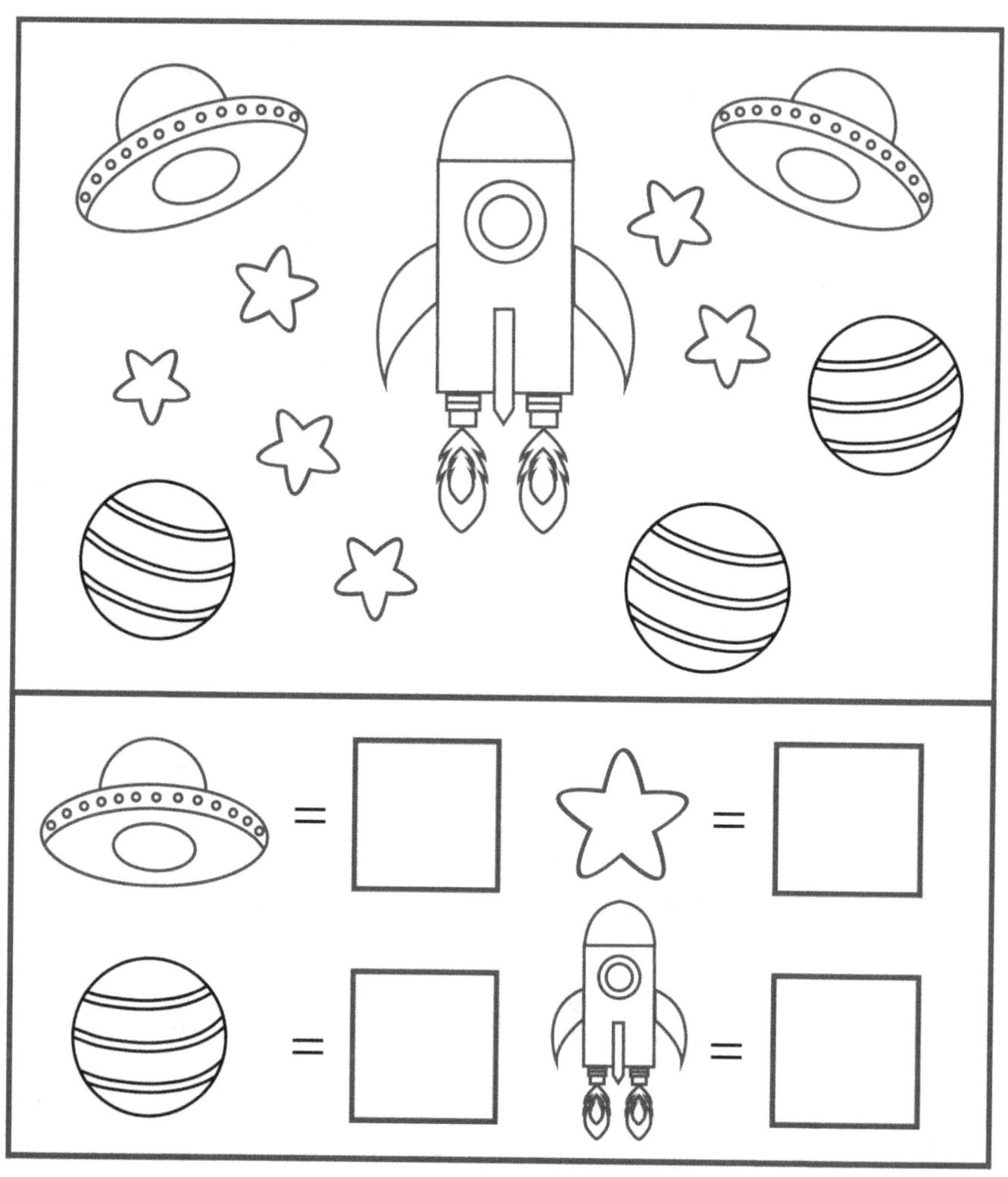

Matching

Add and connect dots with the correct answer

$5 + 2$	●	●	6
$7 + 3$	●	●	9
$4 + 5$	●	●	4
$3 + 5$	●	●	5
$5 + 1$	●	●	7
$2 + 3$	●	●	0
$2 + 2$	●	●	10
$0 + 0$	●	●	8

What comes next?

Draw the next item in the pattern

Subtraction

Subtract the numbers from the middle number

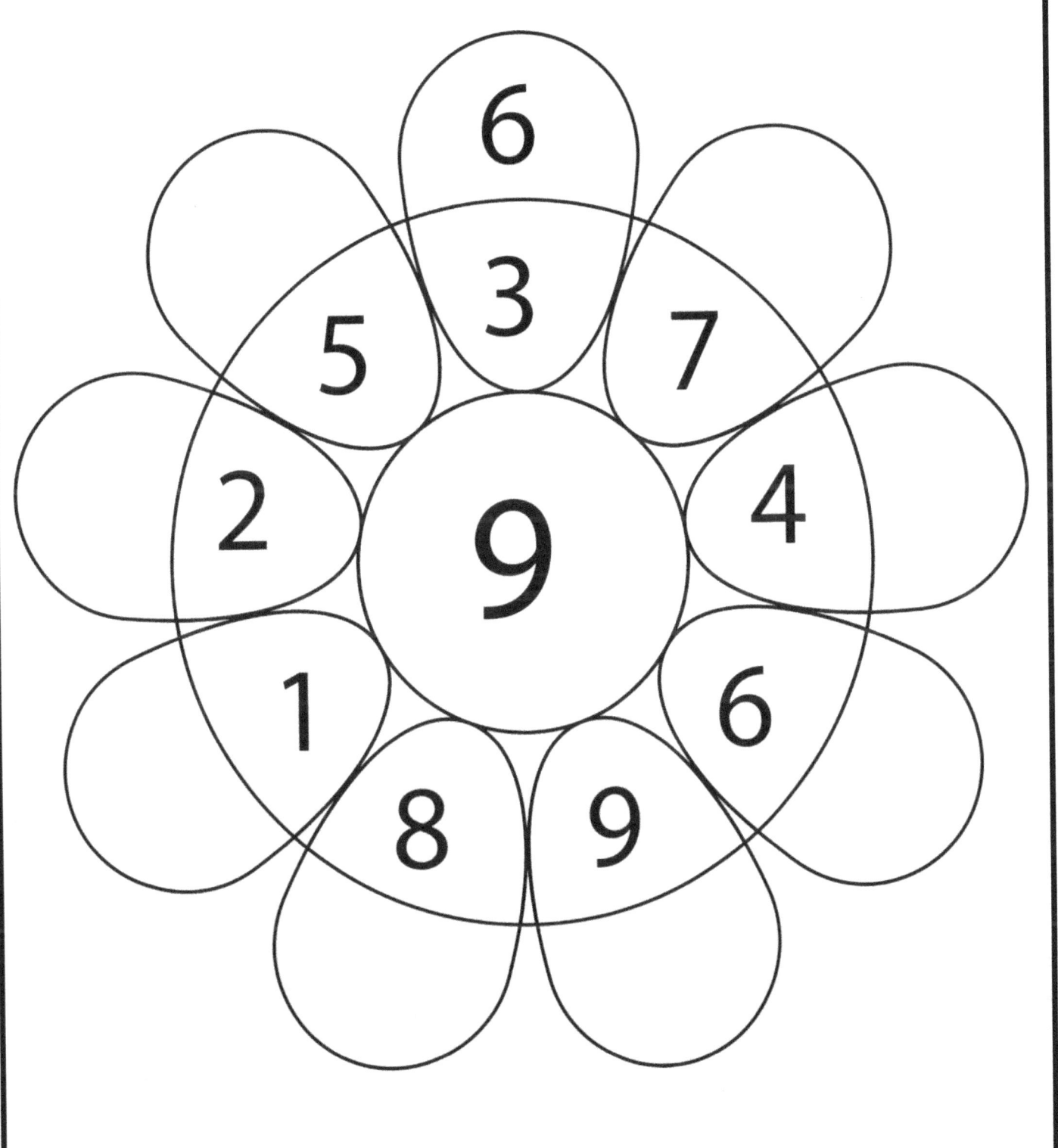

Tracing numbers

Trace the number and count items in each group.
Color the pictures that have 3 items in the set

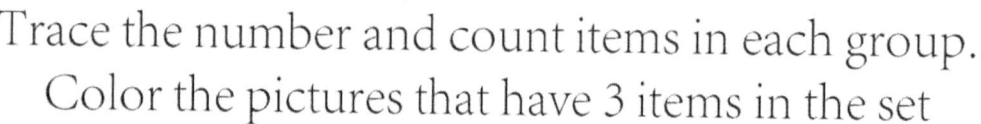

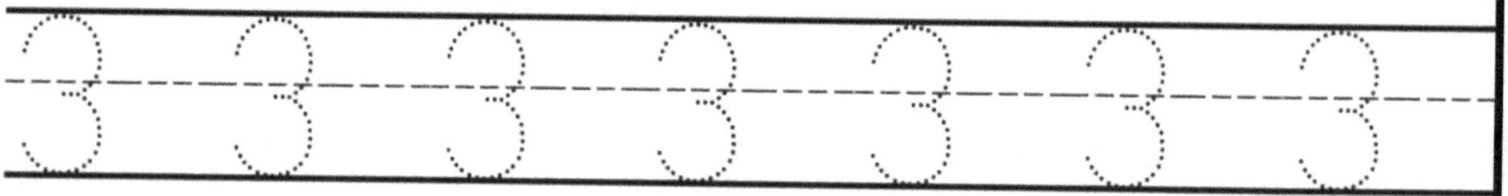

Trace

Trace each number then color the clouds

Number maze

Color the number 4 through the maze to the stop sign

Before and after

Fill in the numbers that come before and after

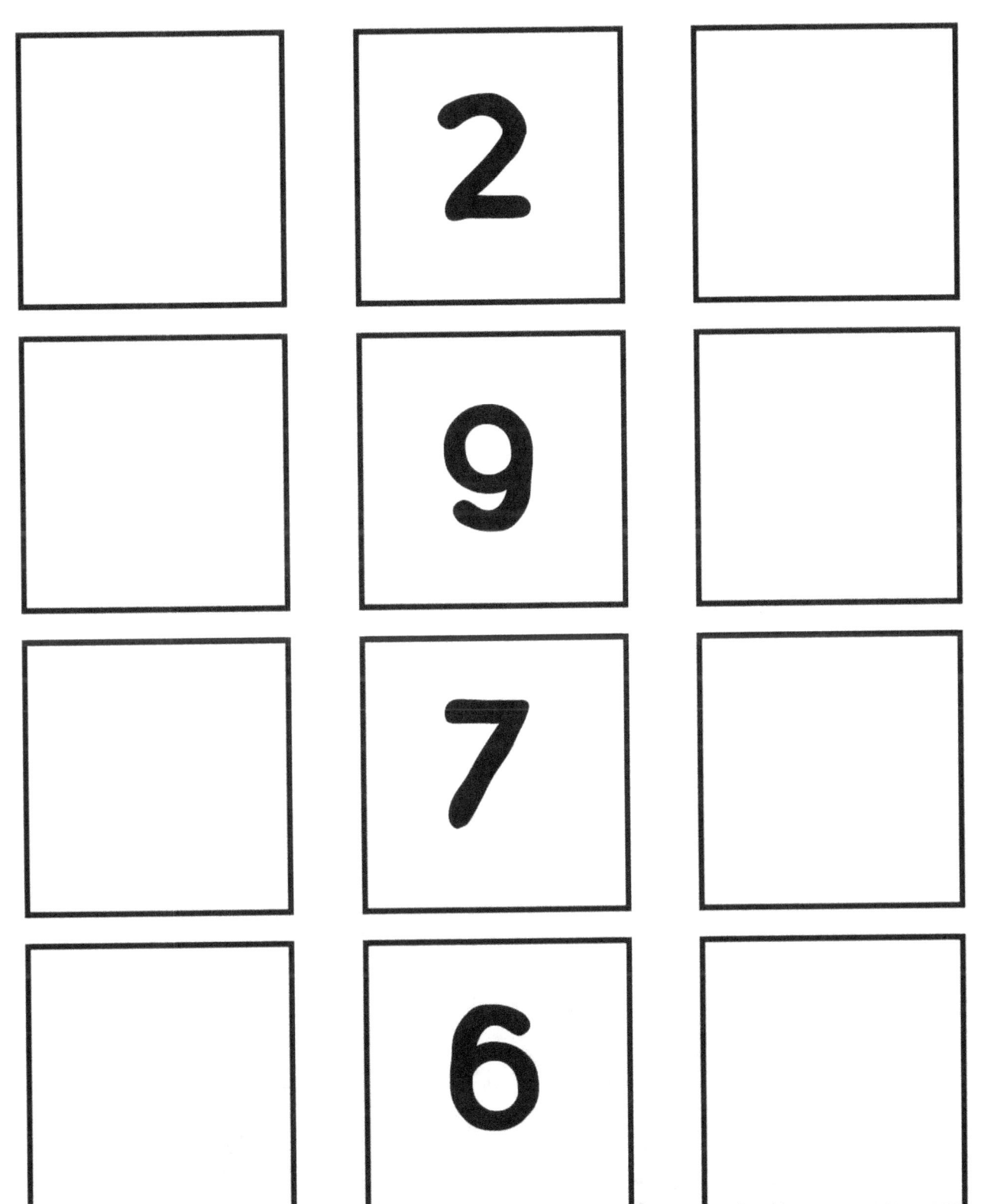

Counting

Count and circle the correct answer

| 4 | 2 | 1 |

| 3 | 10 | 5 |

| 7 | 2 | 8 |

| 7 | 3 | 4 |

Subtraction

Subtract and write your answers in the box then color all the rockets that equal 4

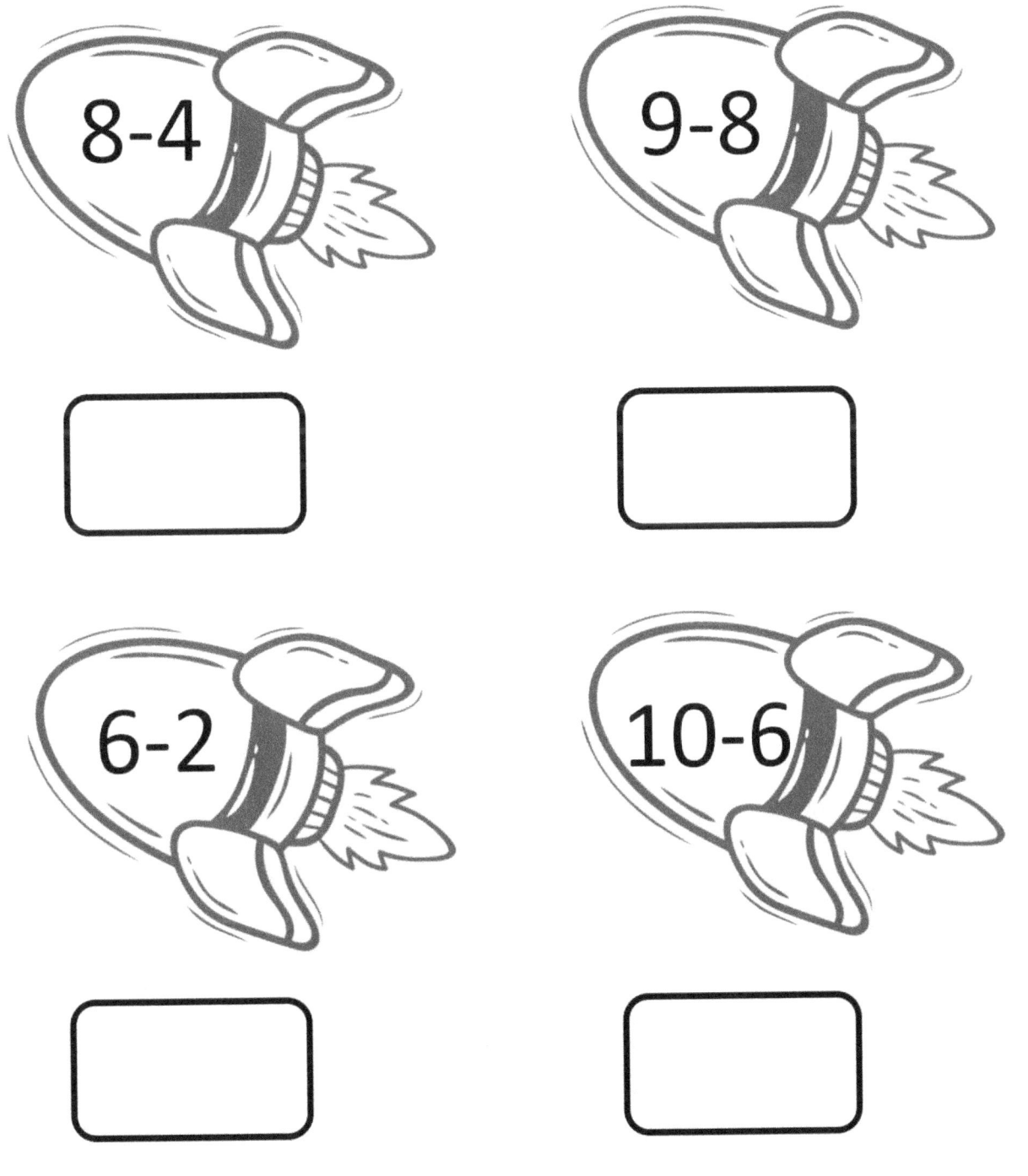

Tell the time

Write the correct time under each clock

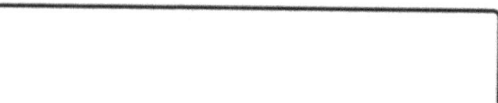

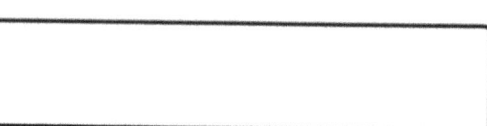

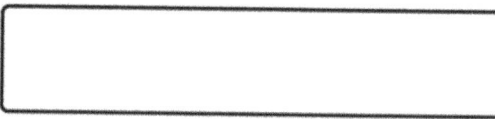

Number maze

Color the number 5 through the maze to the stop sign

Before and after

Fill in the numbers that come before and after

	13	
	14	
	18	
	12	

Addition

Add and write your answers in the
box then color all the peaches that equal 7

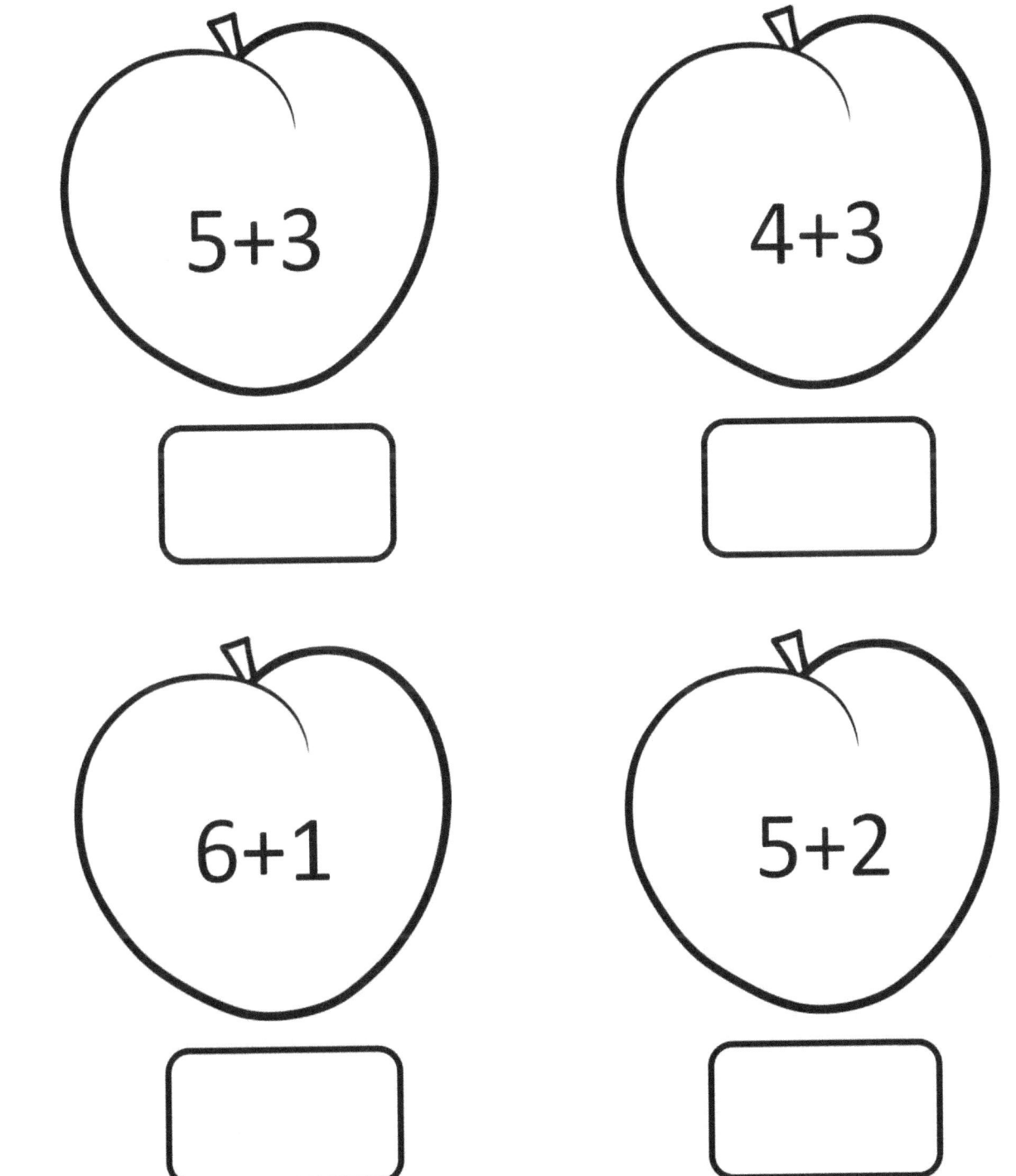

Subtraction

Subtract the shapes and
write the correct answer in the box

   =

  =

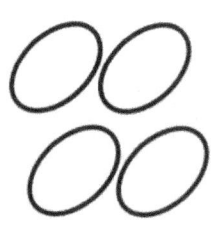  =

Addition

Count and add.
Write the correct answers in the box.

$6 +$

$8 +$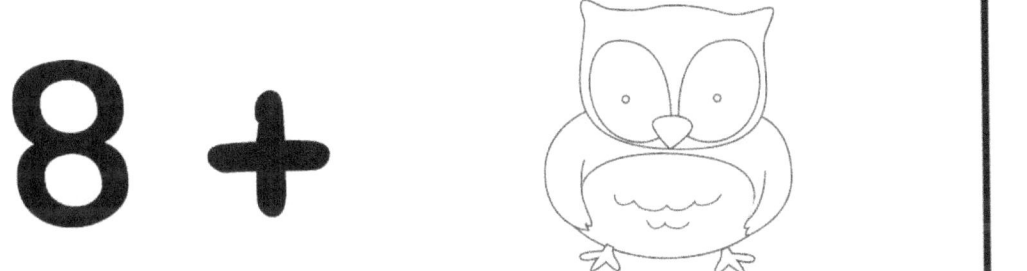

$7 +$

$9 +$

Addition

Count the worms and fill in the boxes

$\boxed{2}$ + 3 = $\boxed{5}$ $\boxed{}$ + 3 = $\boxed{}$

$\boxed{}$ + 3 = $\boxed{}$ $\boxed{}$ + 3 = $\boxed{}$

Tracing numbers

Trace the number and count items in each group.
Color the pictures that have 4 items in the set

4

Counting

Count the objects and write the answers below

Number maze

Color the number 6 through the maze to the stop sign

What comes next?

Draw the next item in the pattern

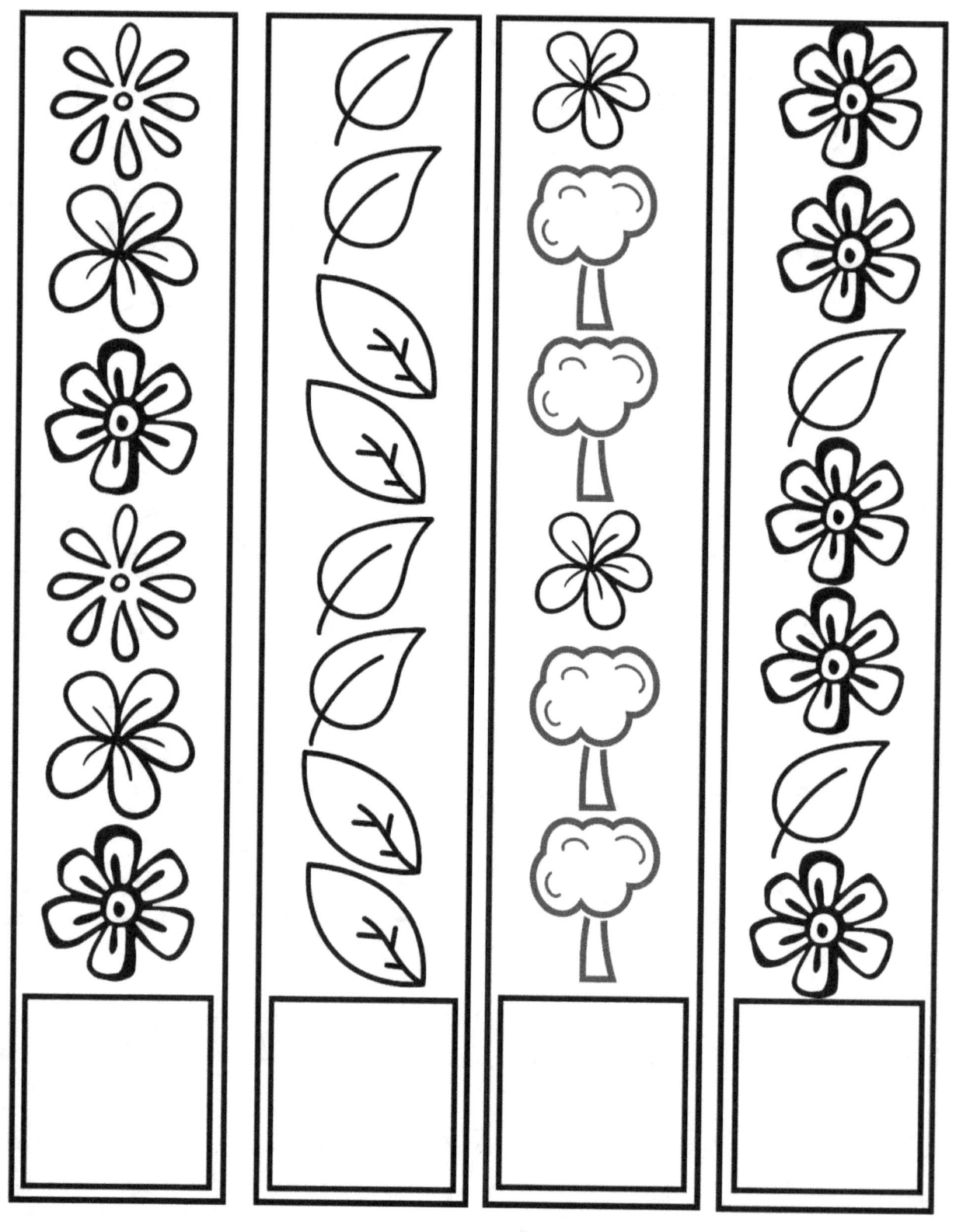

Missing numbers

Find the missing number to complete the sum.

$5 - 2 = \underline{\hspace{2cm}}$

$\underline{\hspace{2cm}} - 2 = 0$

$7 - 3 = \underline{\hspace{2cm}}$

$3 - \underline{\hspace{2cm}} = 1$

$4 - \underline{\hspace{2cm}} = 3$

$\underline{\hspace{2cm}} - 4 = 5$

$8 - \underline{\hspace{2cm}} = 6$

$\underline{\hspace{2cm}} - 1 = 1$

$5 - 3 = \underline{\hspace{2cm}}$

$\underline{\hspace{2cm}} - 3 = 1$

$4 - \underline{\hspace{2cm}} = 2$

$8 - \underline{\hspace{2cm}} = 3$

$9 - 5 = \underline{\hspace{2cm}}$

$6 - \underline{\hspace{2cm}} = 3$

More, less or equal

Choose the correct answer

Addition

Add and write your answers in the
box then color any bugs that equal 7

Addition

Write the correct answer
and mark it on the line

4 + 3 = _____

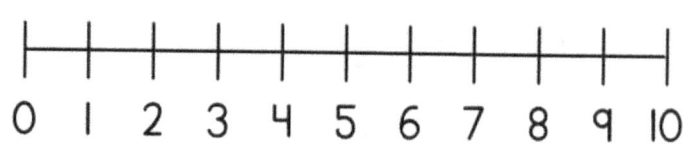

1 + 2 = _____

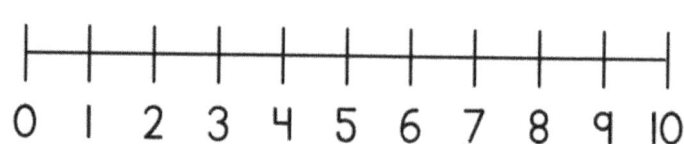

5 + 4 = _____

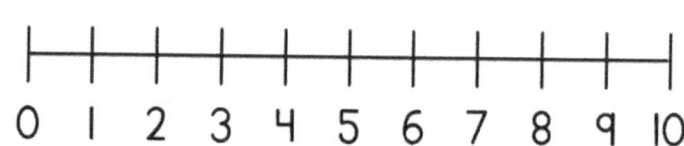

3 + 6 = _____

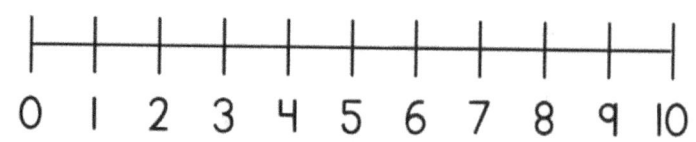

7 + 1 = _____

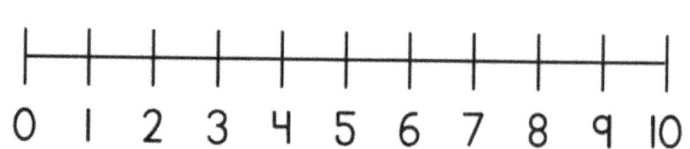

2 + 4 = _____

0 1 2 3 4 5 6 7 8 9 10

More, less or equal

Choose the correct answer

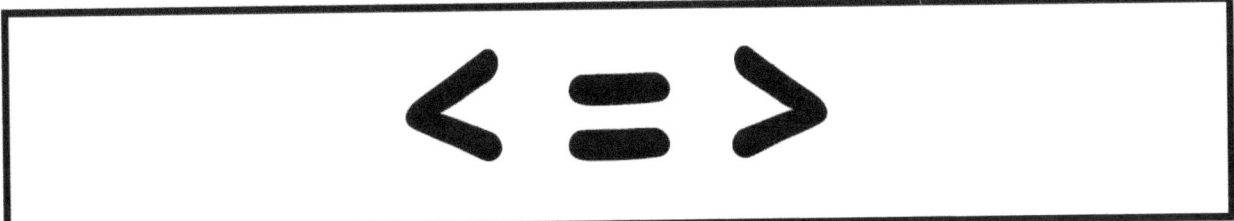

6		7
0		6
9		8

Tell the time

Draw two hands on the
clock face to show the time

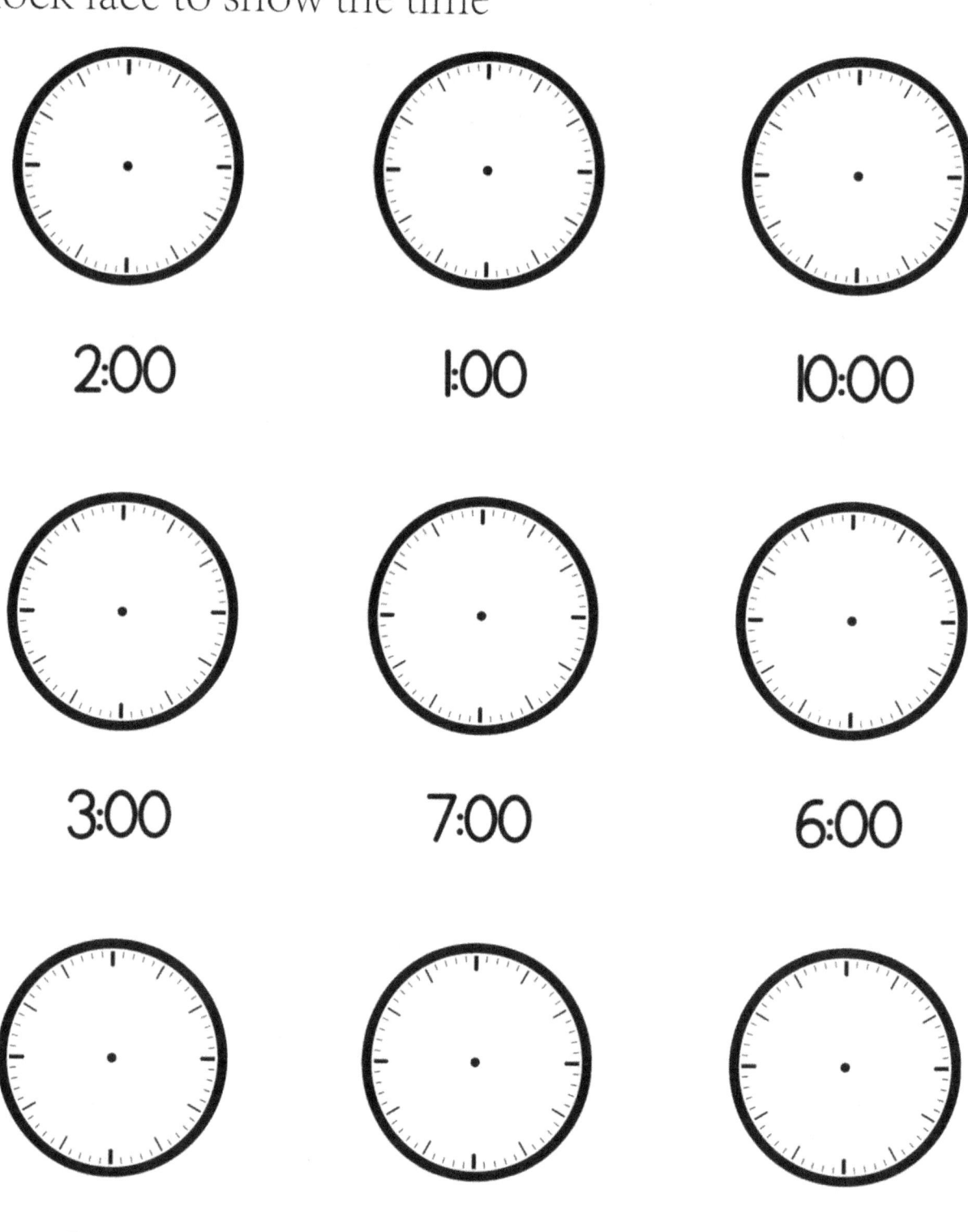

2:00

1:00

10:00

3:00

7:00

6:00

8:00

12:00

11:00

Subtraction

Subtract and write your answers in the
box then color all the strawberries that equal 5

5-5

9-5

8-3

6-1

Counting

Count the objects and write the answers below

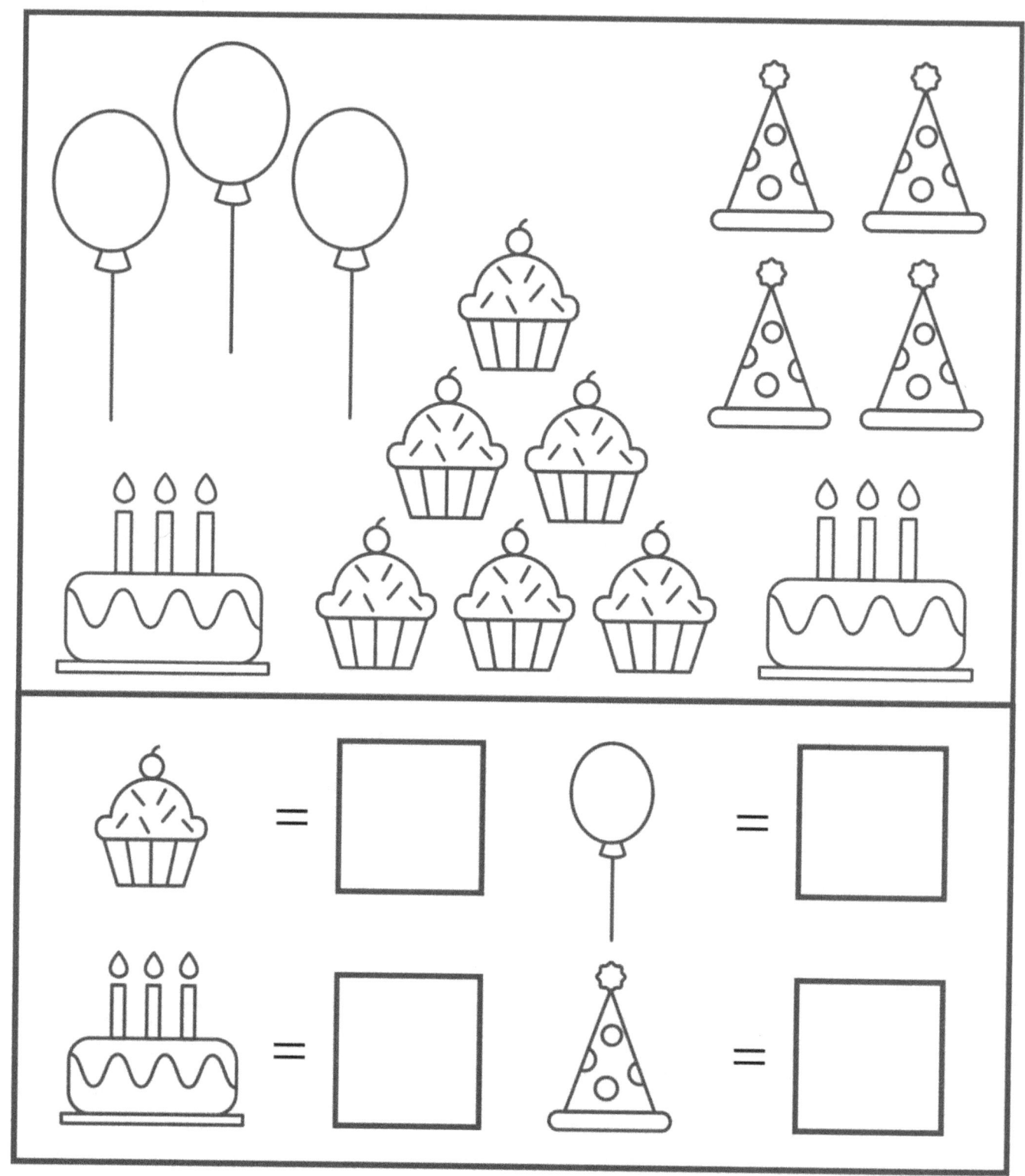

Shapes

Color the shapes

Square	Circle	Rectangle	Triangle
Star	Heart	Rhombus	Oval
Ellipse	Pentagon	Hexagon	Heptagon
Octagon	Nonagon	Parallelogram	Trapezium
Kite	Cross	Arrow	Crescent

Tracing numbers

Trace the number and count items in each group.
Color the pictures that have 5 items in the set

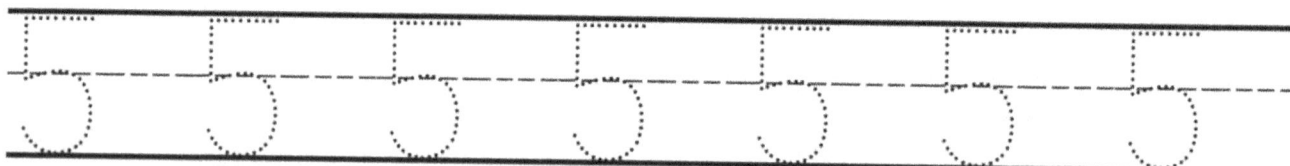

Addition

Write the correct answer
and mark it on the line

5 + 1 = _____
_____

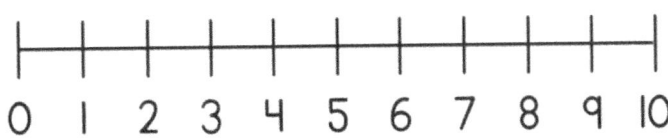

2 + 4 = _____
_____

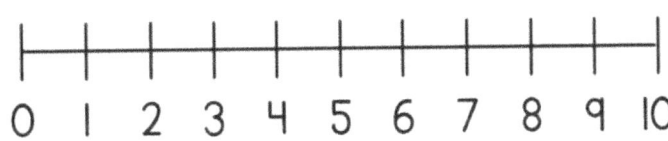

3 + 1 = _____
_____

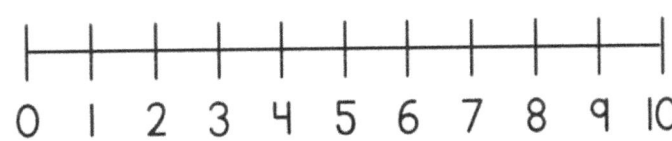

6 + 2 = _____
_____

7 + 2 = _____
_____

1 + 8 = _____
_____

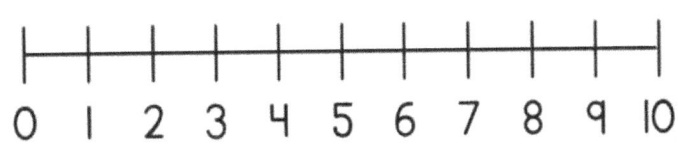

Addition

Count and add.
Write the correct answers in the box.

Missing numbers

Fill the missing numbers below

1		3	4		6			9	
	12			15		17	18		20
		23	24		26		28		
31	32			35		37		39	40
		43	44		46	47			50
51			55	56		58			
	62	63				67		69	70
71	72		75			78			
		83	84		86	87			90
91	92		95		97			99	100

Addition

Count and add. Write the correct answers in the box.

4 + =

6 + =

2 + =

3 + =

5 + =

Count and color

Color the boxes to match
the number on the fish

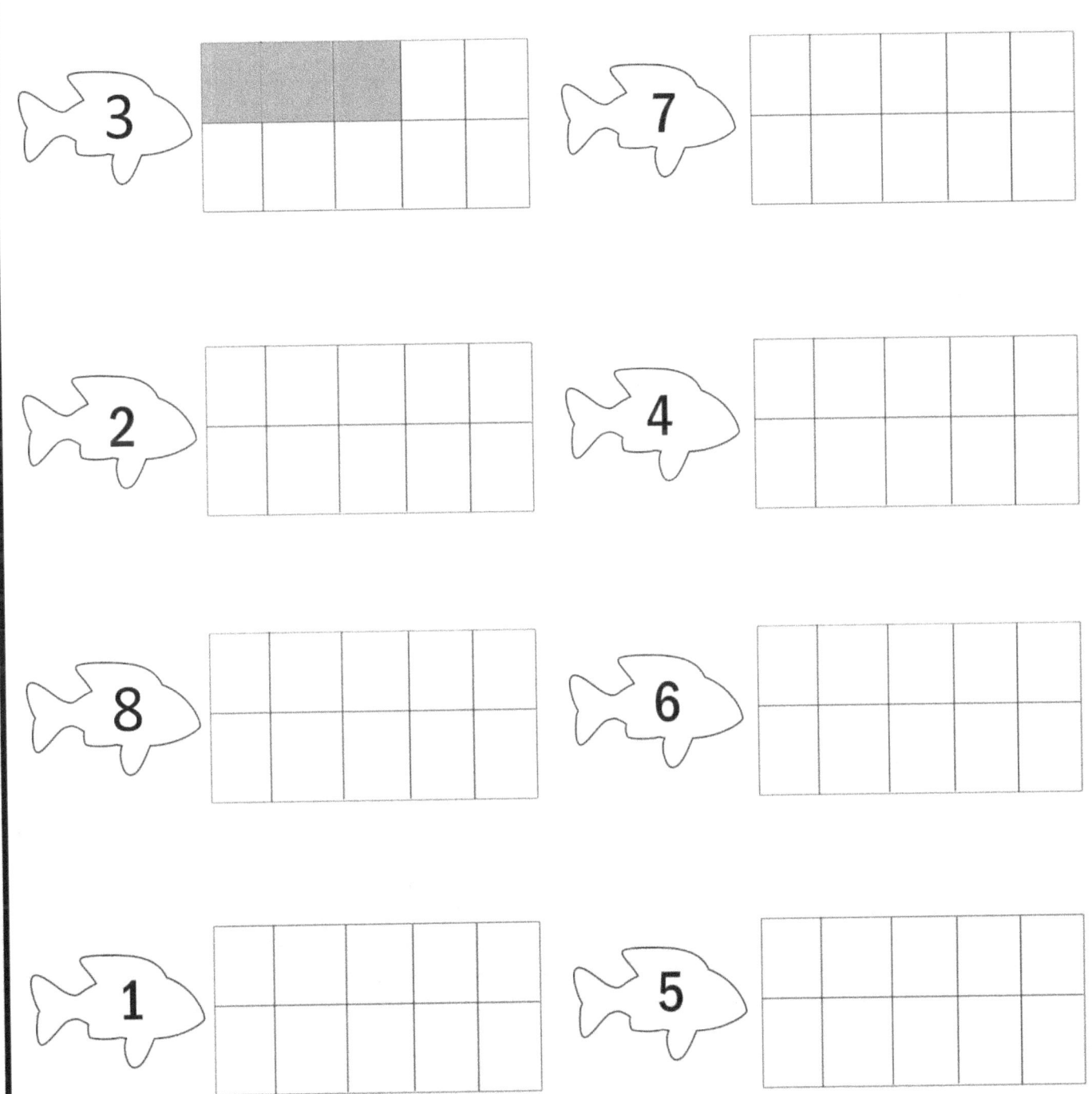

Tracing numbers

Trace the number and count items in each group.
Color the pictures that have 6 items in the set

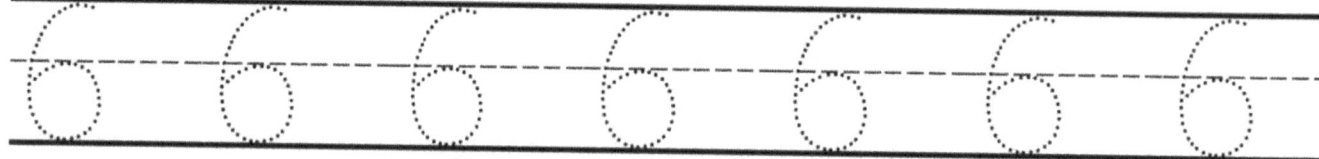

Addition

Add and write your answers in the
box then color all the hippos that equal 9

Number maze

Color the number 7 through the maze to the stop sign

Count and color

How many unicorns?
Circle the correct number

3   4   2

Drawing

Draw a group with **<u>more</u>** hearts.

Draw a group with **<u>fewer</u>** circles.

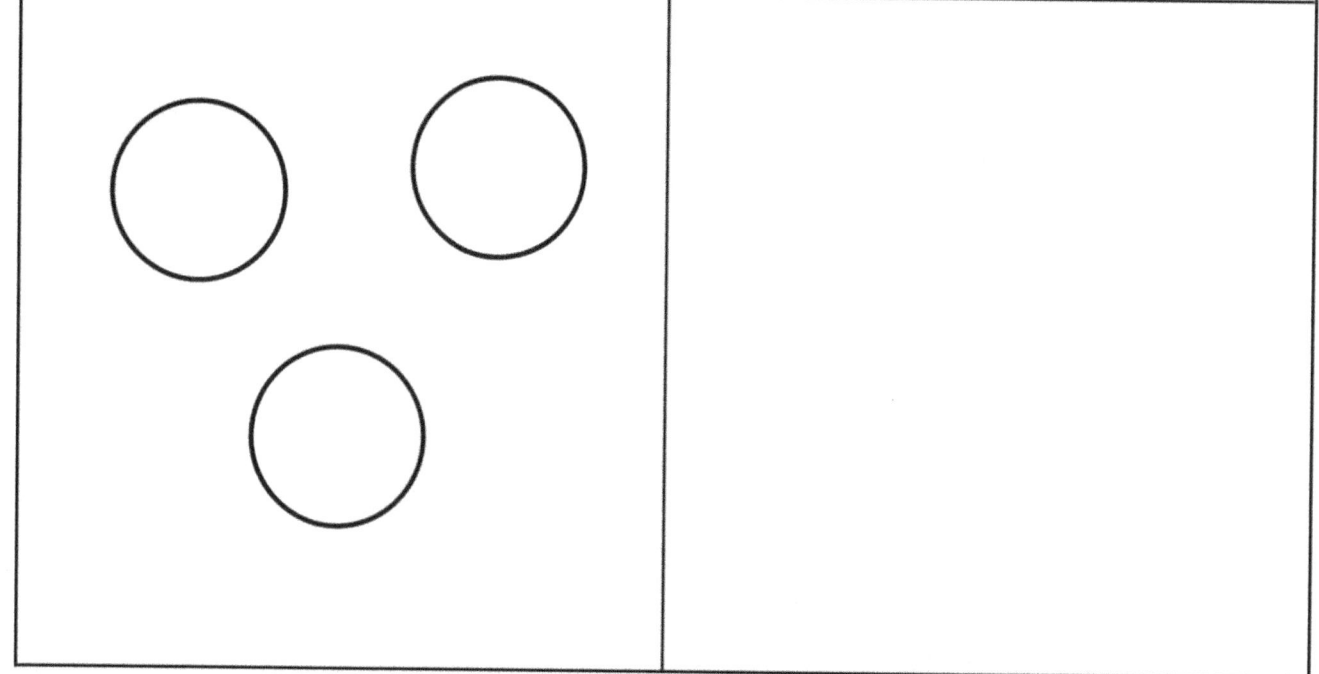

Addition

Add and write your answers in the tree

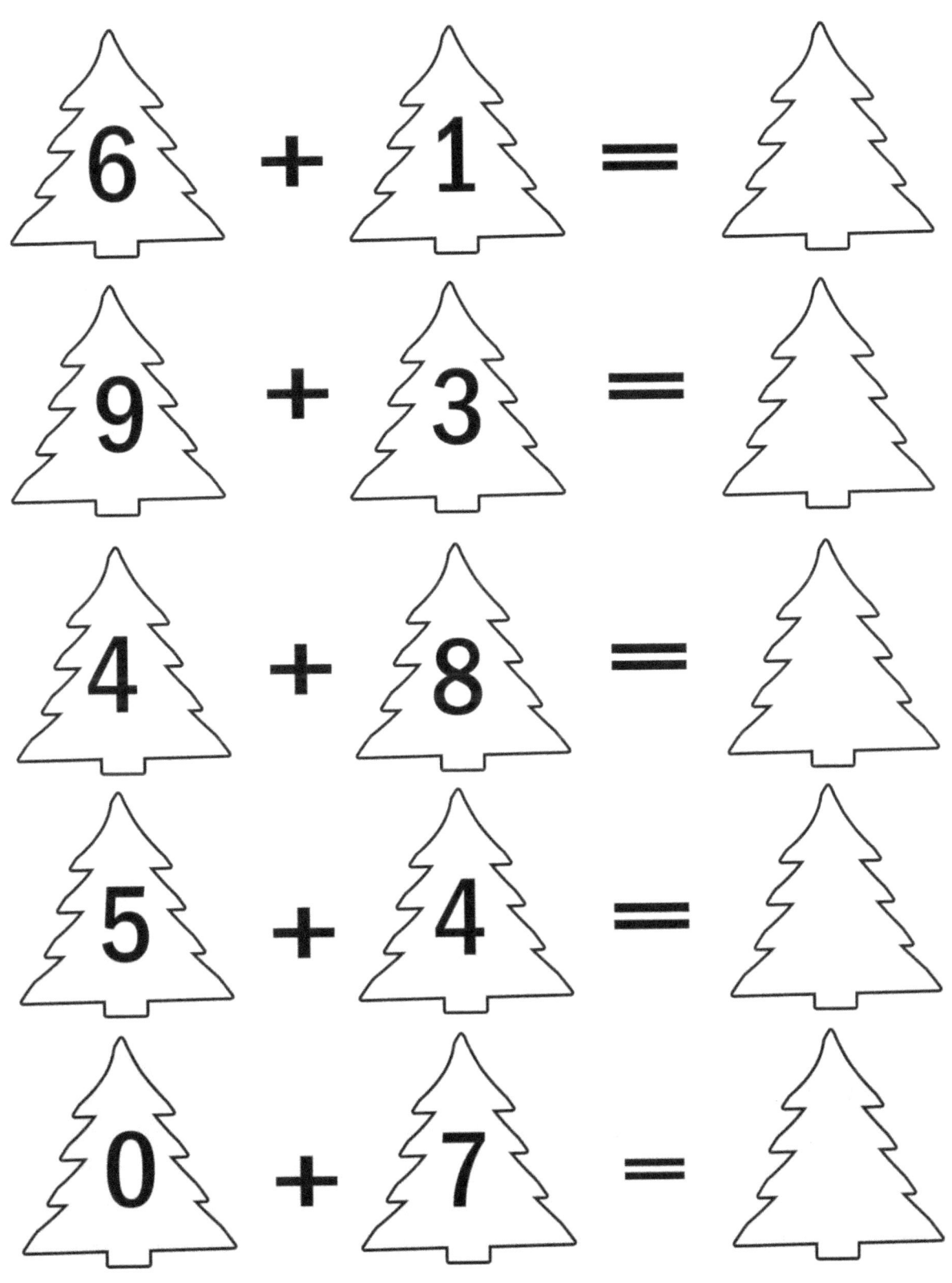

6 + 1 =

9 + 3 =

4 + 8 =

5 + 4 =

0 + 7 =

More, less or equal

Choose the correct answer

9		9
10		8
8		8

Shapes

Trace the shapes

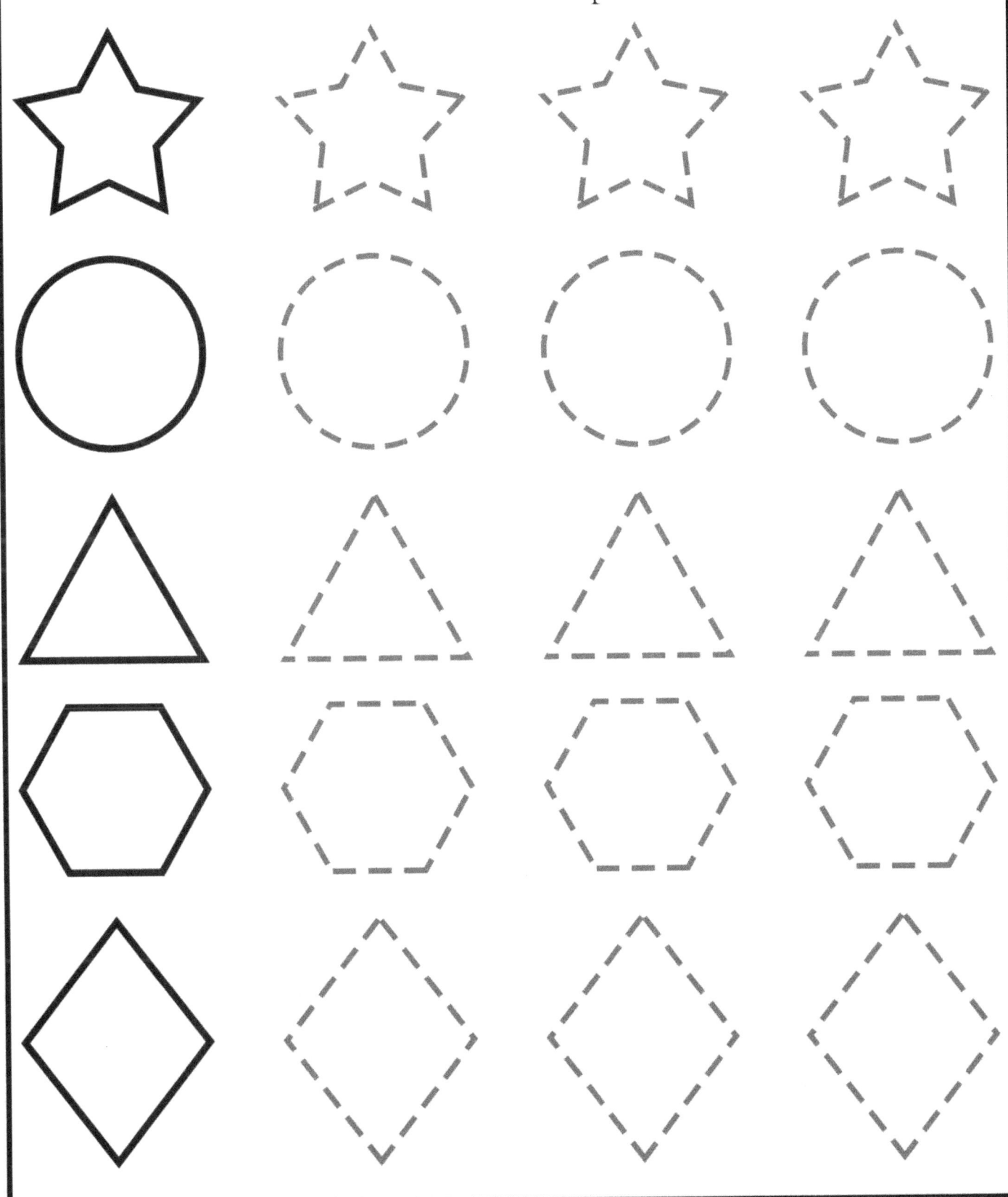

Counting

Count the objects and write the answers below

Drawing

Draw a group with **<u>more</u>** circles.

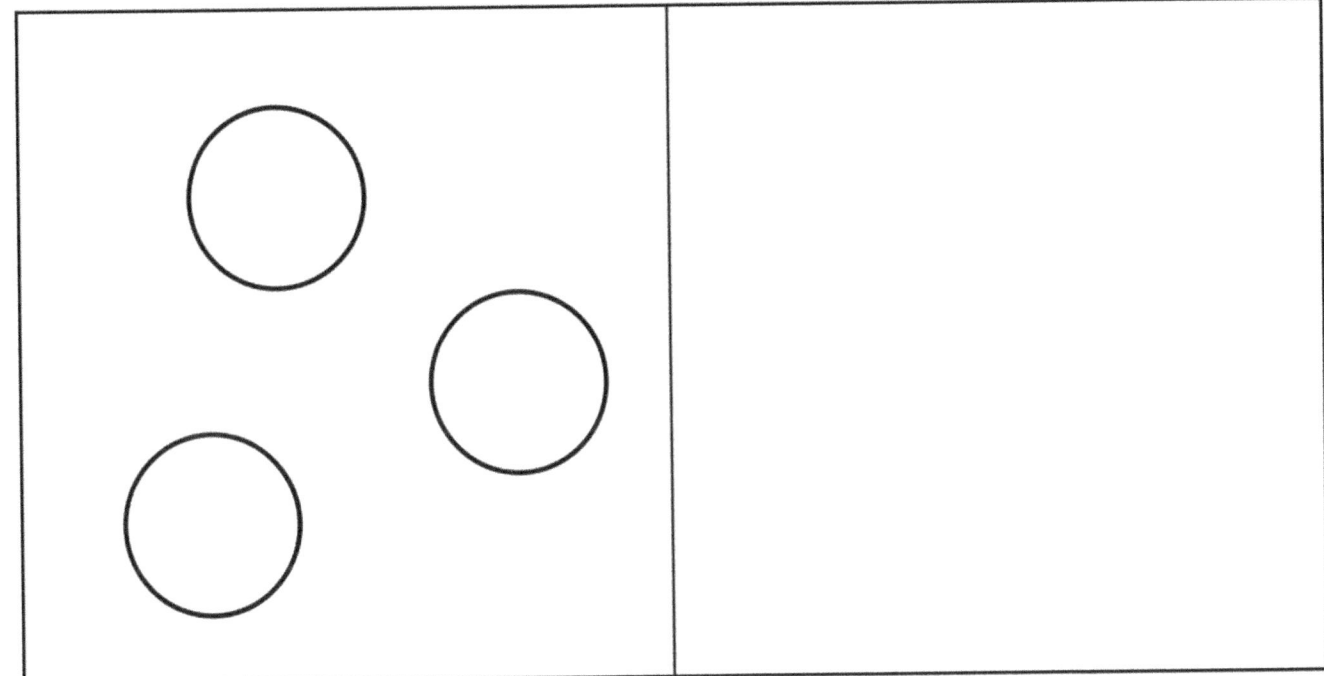

Draw a group with **<u>fewer</u>** triangles.

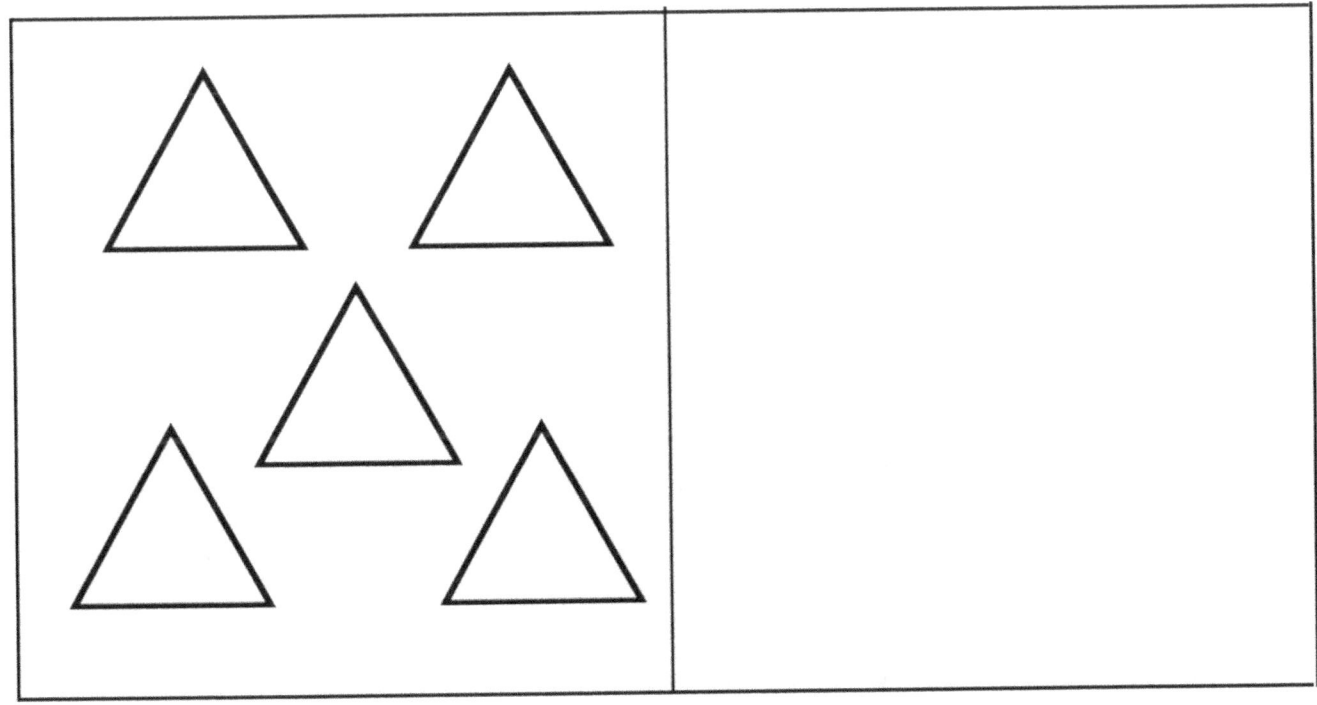

Tracing numbers

Trace the number and count items in each group.
Color the pictures that have 7 items in the set

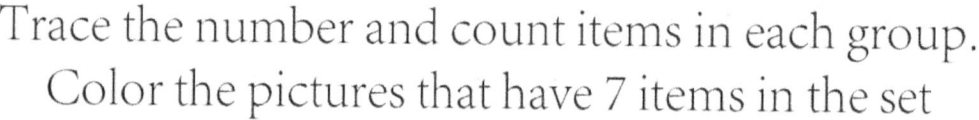

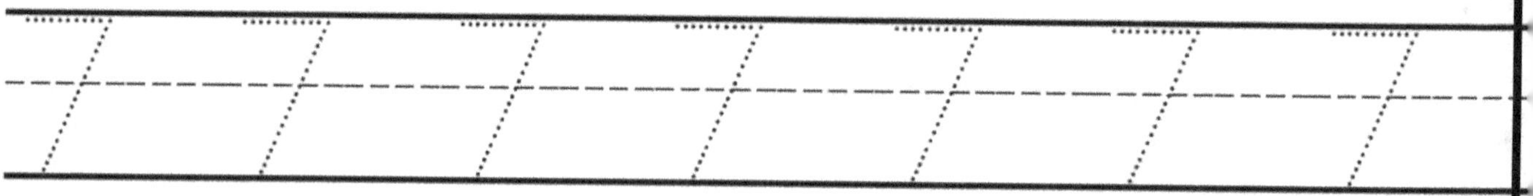

Number maze

Color the number 8 through the maze to the stop sign

Before and after

Fill in the numbers that come before and after

Comparing

Color the numbers that have the bigger value.

5 8	4 5
6 0	3 2
9 5	4 7
8 2	6 1
3 7	5 2

Subtraction

Count and subtract.
Write the correct answers in the box.

5 -

3 -

4 -

2 -

Addition

Count and add. Write the correct answers in the box.

4 + 🐘🐘🐘🐘 = ☐

6 + 🦉 = ☐

7 + 🐦🐦 = ☐

2 + 🦘🦘🦘 = ☐

1 + 🦛🦛 = ☐

Count and color

How many children?
Circle the correct number

2   4   3

Before and after

Fill in the numbers that come before and after

	11	
	17	
	16	
	19	

Subtraction

Subtract the shapes and
write the correct answer in the box

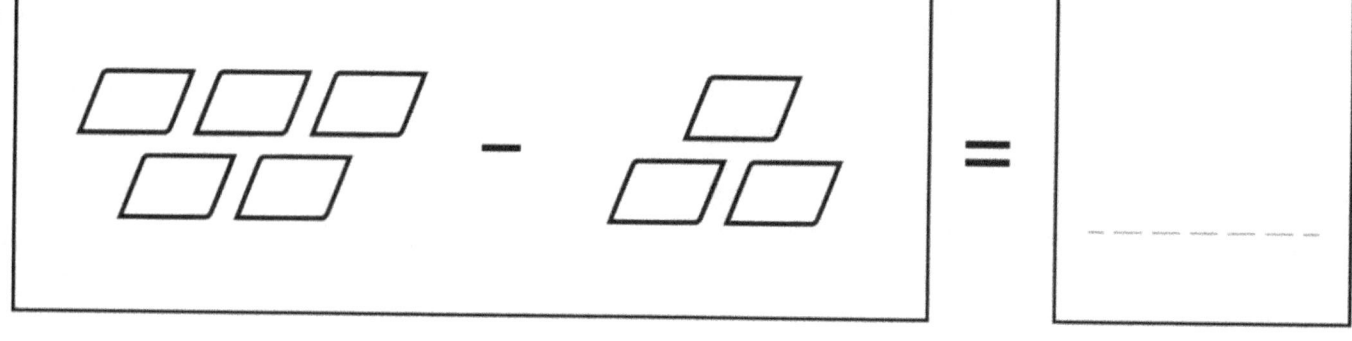

=

=

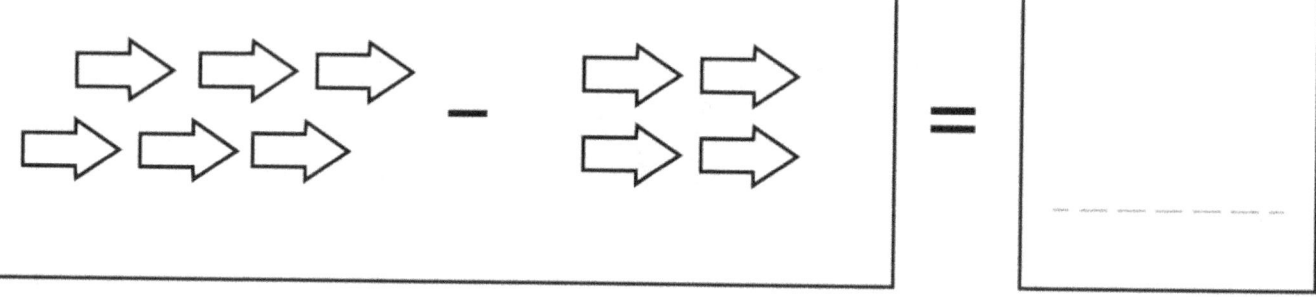

=

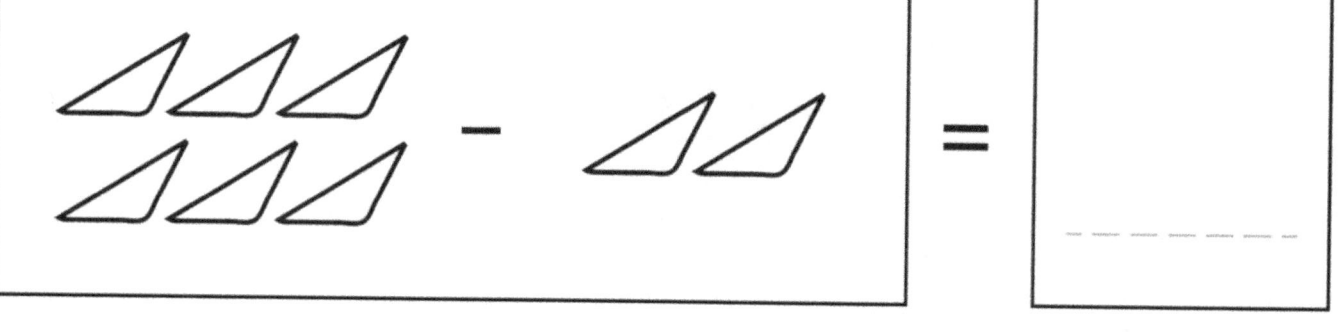

=

Count and color

Color the boxes to match
the number on the starfish

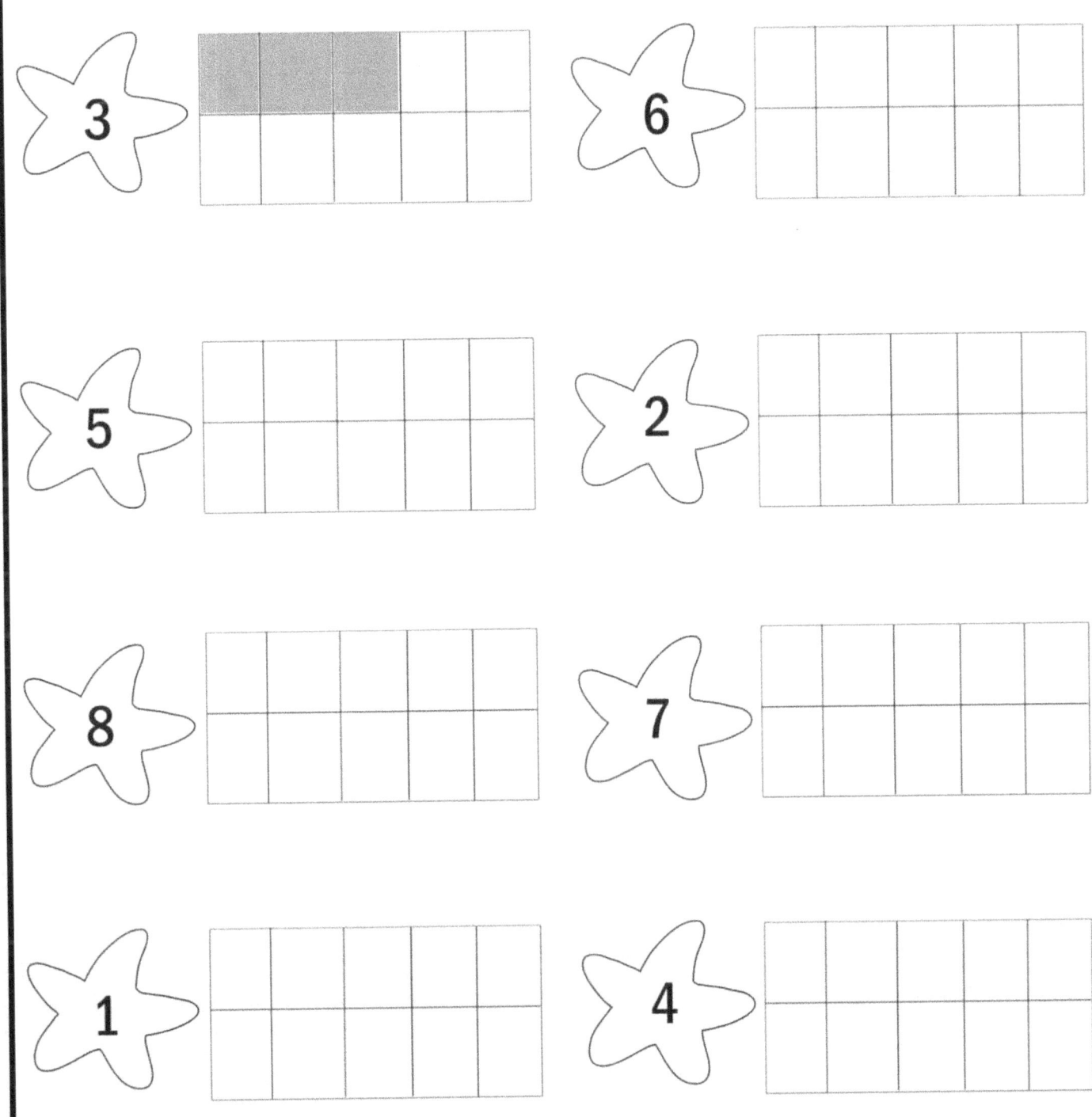

Drawing

Draw a group with <u>**more**</u> squares.

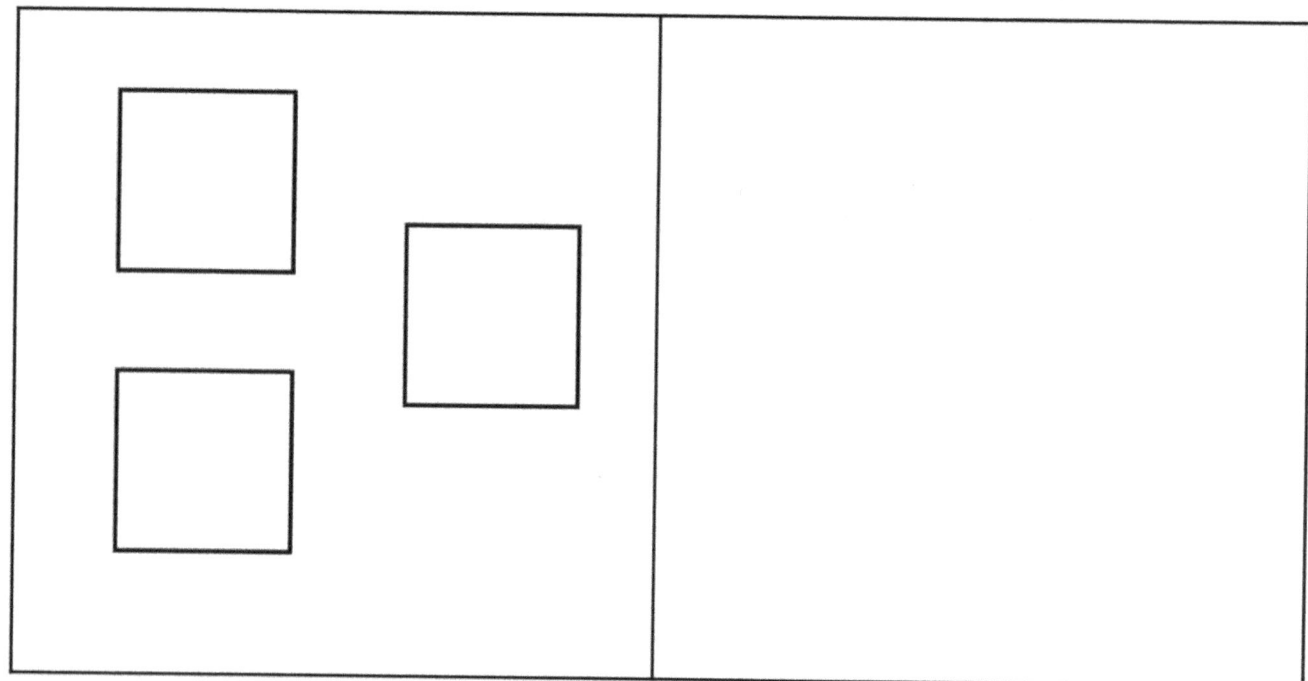

Draw a group with <u>**fewer**</u> stars.

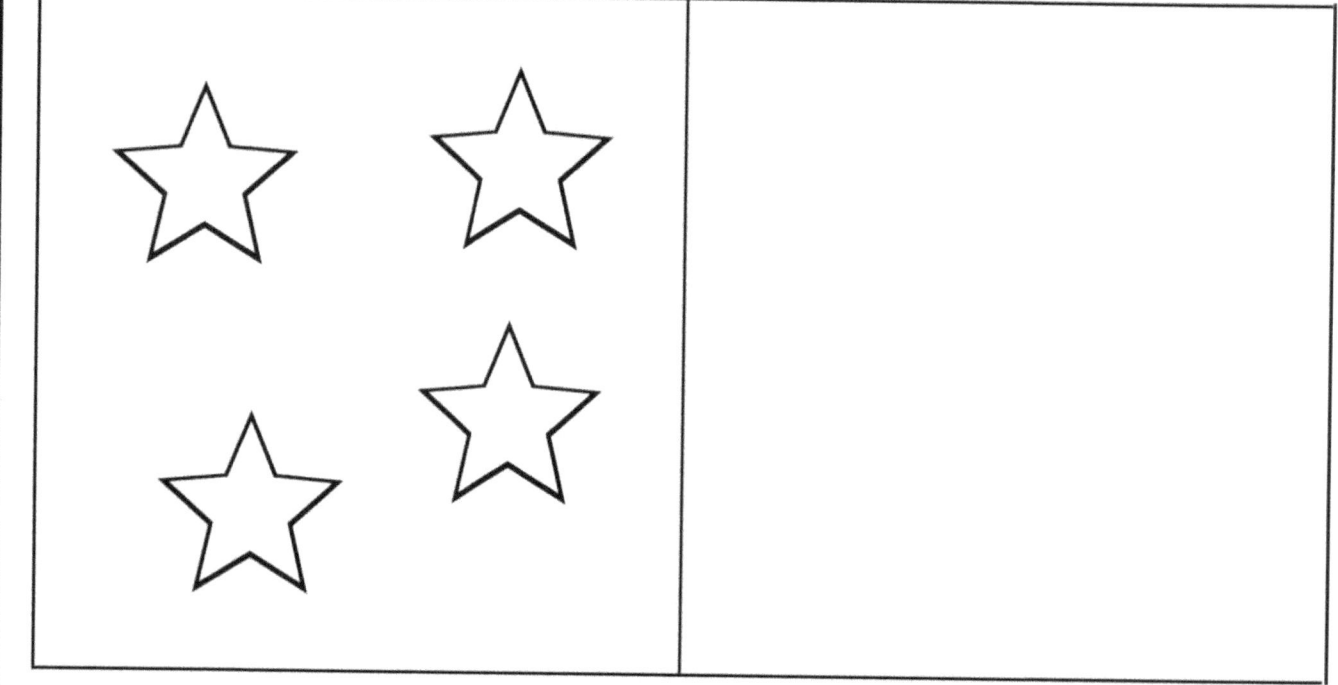

Counting

How many watermelons?

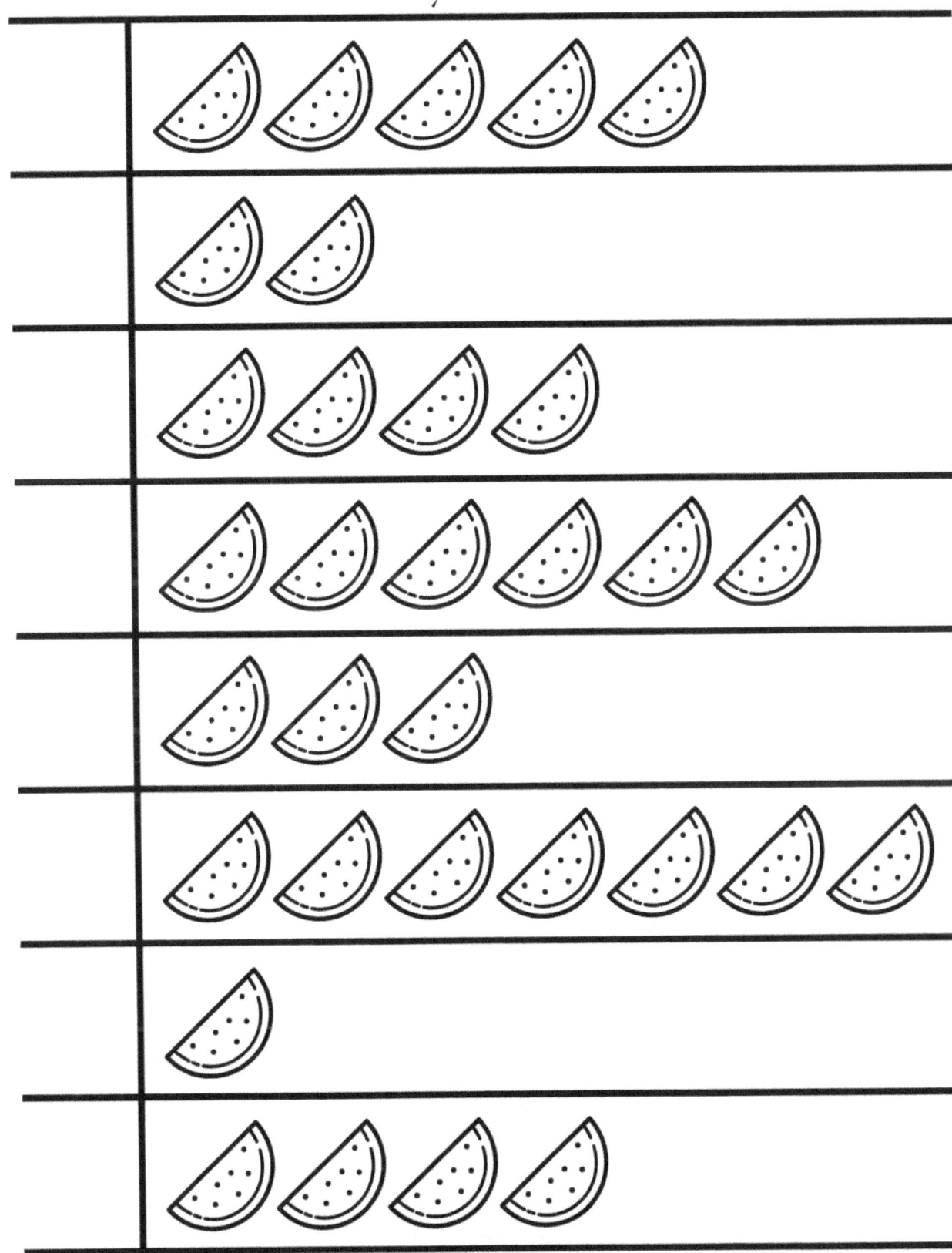

Tracing numbers

Trace the number and count items in each group.
Color the pictures that have 8 items in the set

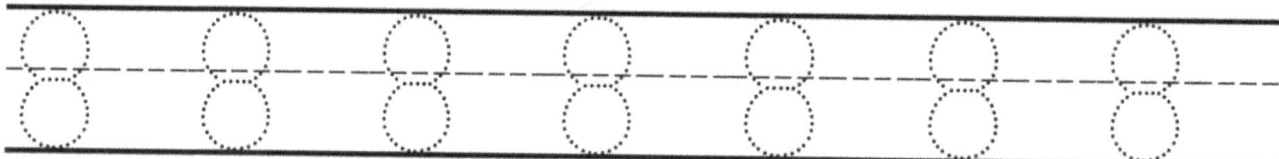

Counting

Color the correct number of shapes.

4

2

5

1

3

Comparing

Color the numbers that have the smaller value.

6 8	4 3
2 0	3 6
7 5	4 9
1 2	6 8
2 7	5 3

Shapes

Trace the shapes

Number maze

Color the number 9 through the maze to the stop sign

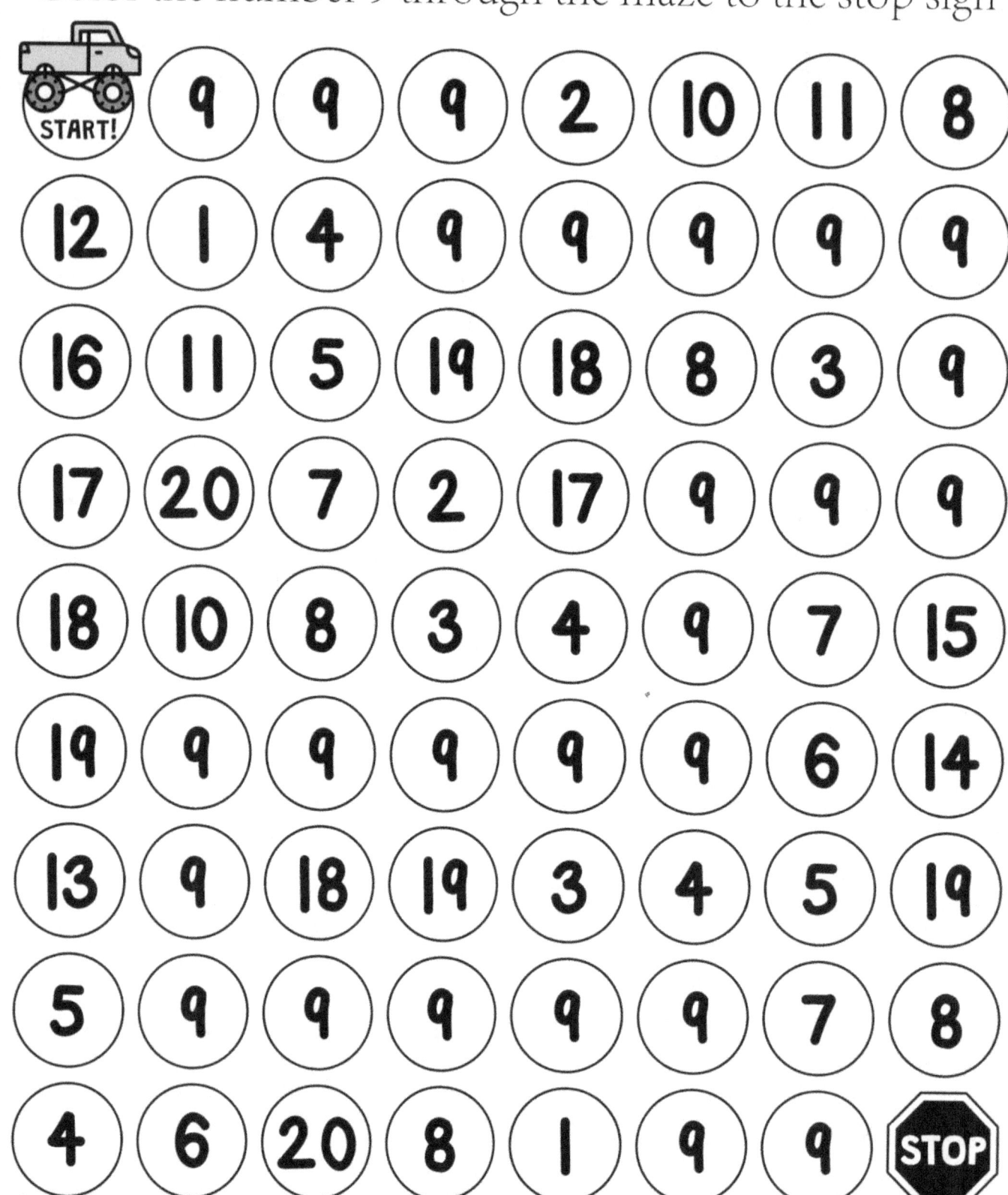

Tracing numbers

Trace the number and count items in each group.
Color the pictures that have 9 items in the set

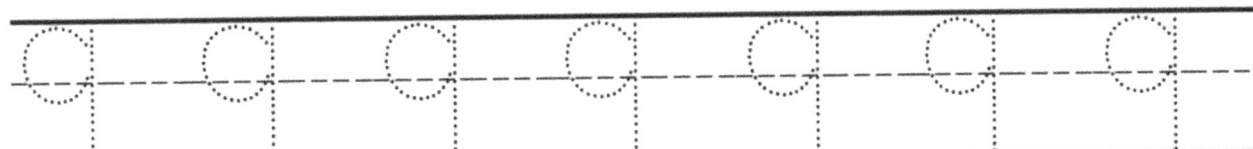

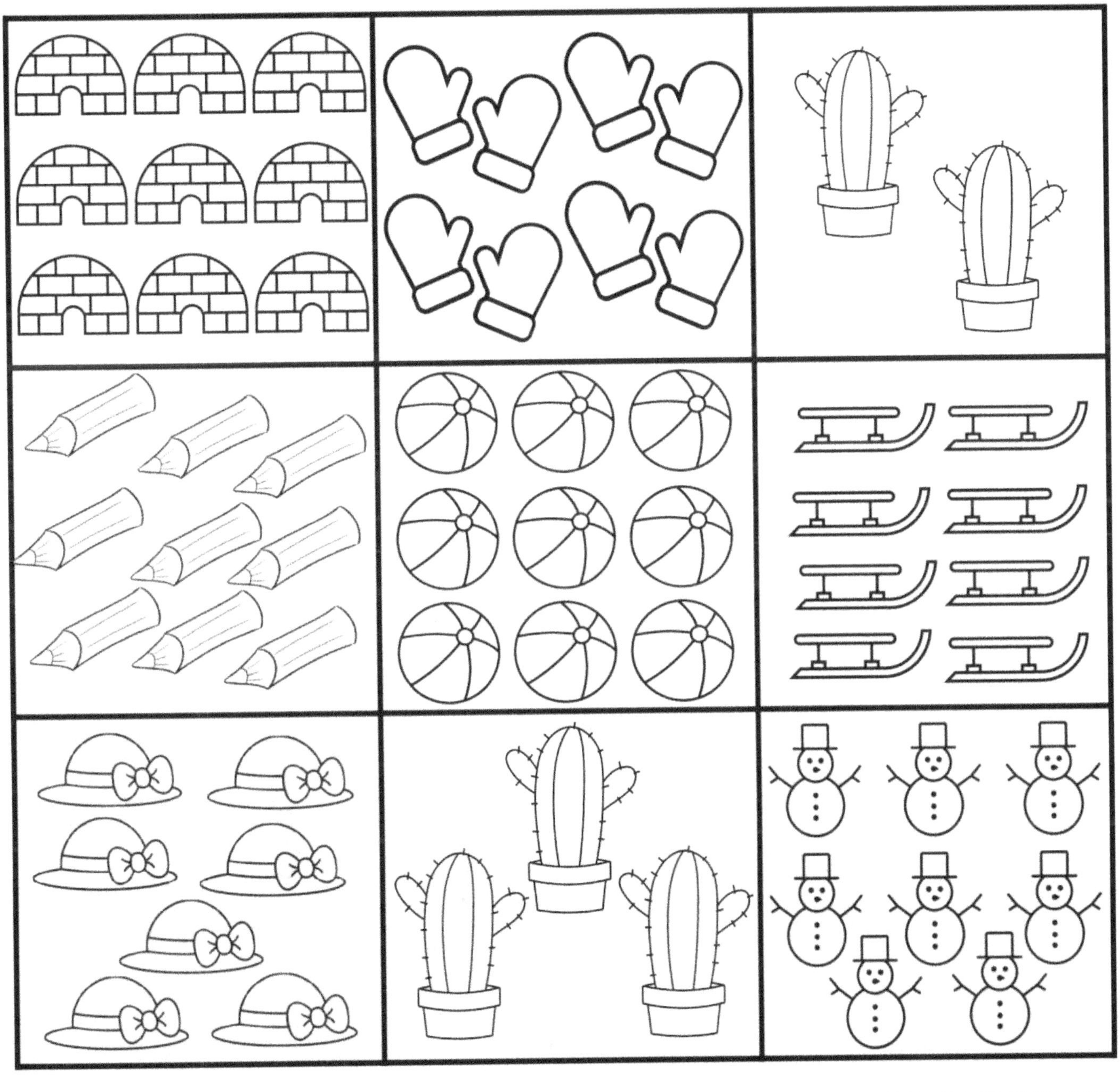

9 9 9 9 9 9 9

Tell the time

Draw two hands on the
clock face to show the time

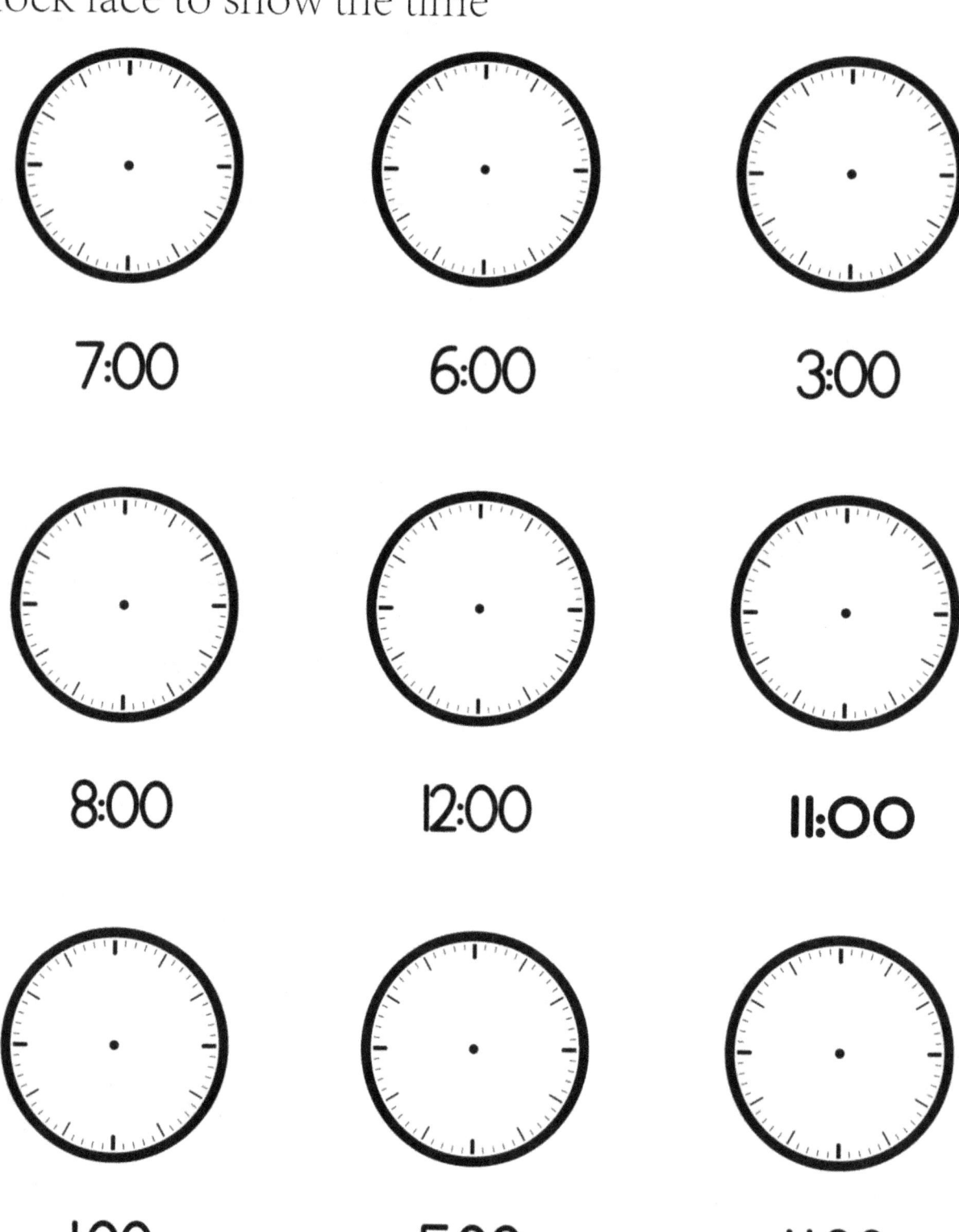

7:00 6:00 3:00

8:00 12:00 11:00

1:00 5:00 4:00

Addition

Write the correct answer
and mark it on the line

6 + 2 = _____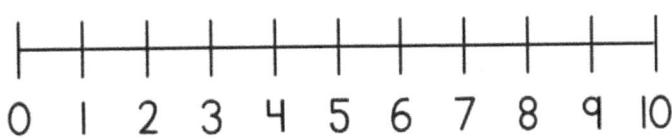

3 + 5 = _____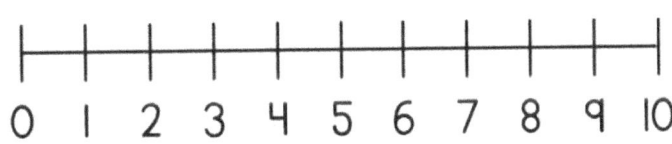

4 + 2 = _____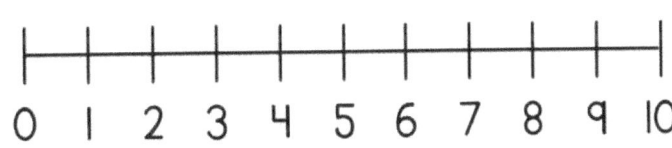

7 + 1 = _____

           0 1 2 3 4 5 6 7 8 9 10

8 + 0 = _____

           0 1 2 3 4 5 6 7 8 9 10

0 + 9 = _____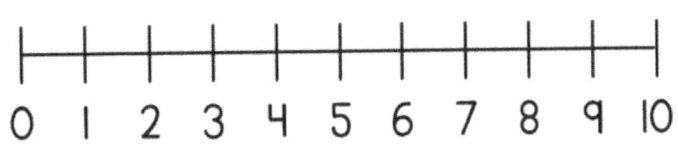

Number maze

Color the number 10 through the maze to the stop sign

Subtraction

Subtract the shapes and
write the correct answer in the box

 =

 =

 =

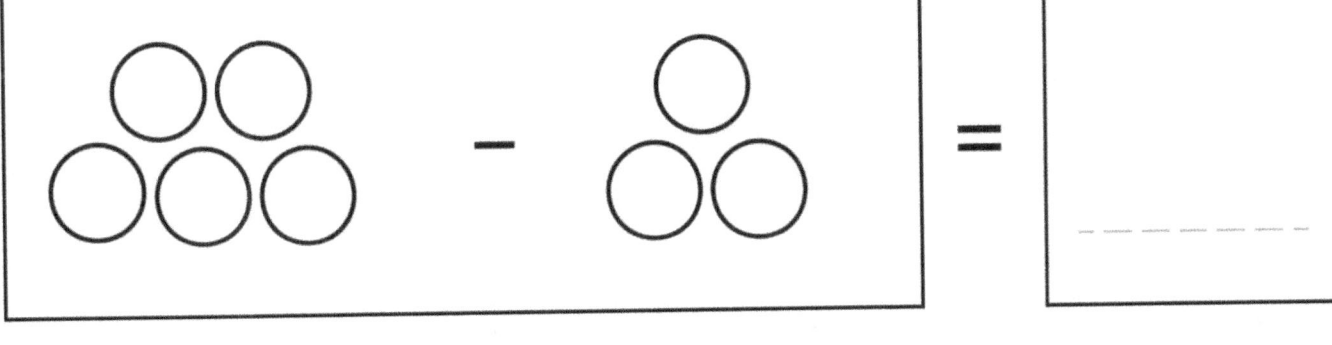 =

Tracing numbers

Trace the number and count items in each group.
Color the pictures that have 10 items in the set

10

10 10 10 10 10 10 10

More, less or equal

Choose the correct answer

3		**5**
2		**1**
3		**3**

Counting

Count the objects and write the answers below

Tracing

Trace each shape and color it

Count and color

Color the boxes to match the number on the Santa

Count and color

Count and circle the correct answer

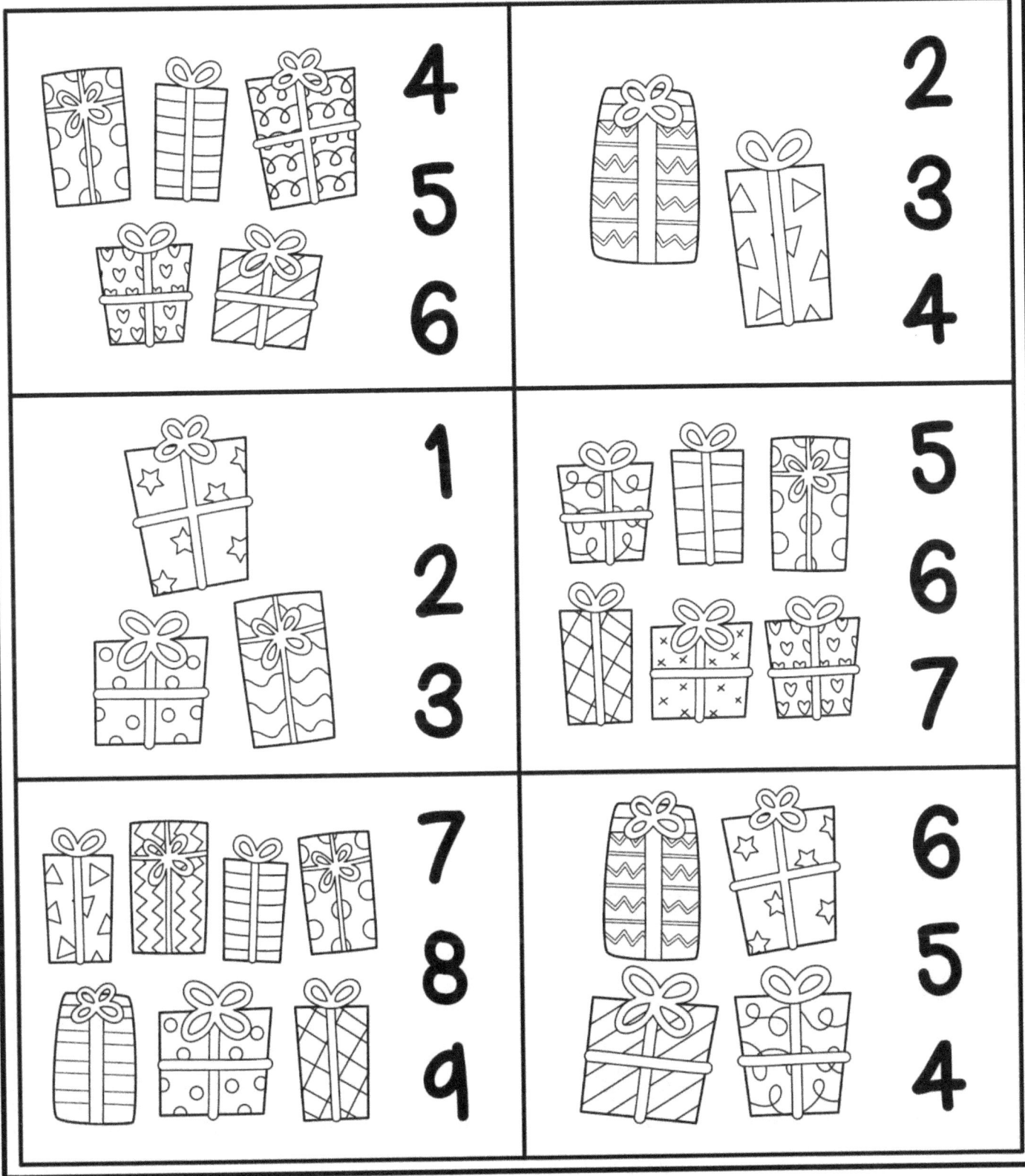

Missing numbers

Find the missing number to complete the sum.

_____ $- 1 = 4$

$6 -$ _____ $= 5$

$4 -$ _____ $= 3$

_____ $- 3 = 1$

_____ $- 1 = 1$

_____ $- 2 = 3$

_____ $- 2 = 4$

_____ $- 4 = 4$

_____ $- 3 = 4$

_____ $- 4 = 3$

$5 -$ _____ $= 2$

_____ $- 1 = 2$

$8 -$ _____ $= 3$

$9 -$ _____ $= 5$

Subtraction

Count and subtract.
Write the correct answers in the box.

9 -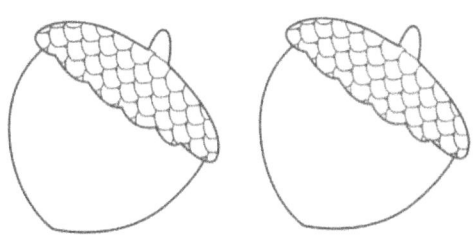

8 - 🌰🌰🌰🌰

7 - 🌰🌰🌰

6 -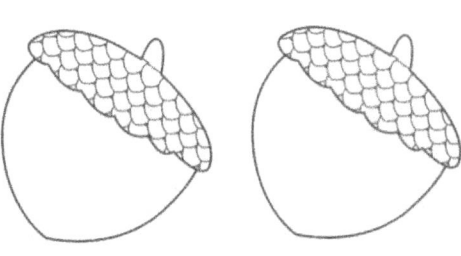

More, less or equal

Choose the correct answer

< = >

4		11
11		12
7		7